Low Electromagnetic Field Exposure Wireless Devices

Low Electromagnetic Field Exposure Wireless Devices

Fundamentals and Recent Advances

Edited by

Masood Ur Rehman
University of Glasgow, Glasgow, UK

Muhammad Ali Jamshed
University of Glasgow, Glasgow, UK

IEEE PRESS
WILEY

Published by John Wiley & Sons, Inc., Hoboken, New Jersey.
Published simultaneously in Canada.

For general information on our other products and services or for technical support, please contact our Customer Care Department within the United States at (800) 762-2974, outside the United States at (317) 572-3993 or fax (317) 572-4002.

Wiley also publishes its books in a variety of electronic formats. Some content that appears in print may not be available in electronic formats. For more information about Wiley products, visit our web site at www.wiley.com.

Library of Congress Cataloging-in-Publication Data Applied for:
Hardback ISBN: 9781119909163

Cover Design: Wiley
Cover Image: © whiteMocca/Shutterstock

Set in 9.5/12.5pt STIXTwoText by Straive, Chennai, India

I dedicate this effort to my parents, Khalil Ur Rehman and Ilfaz Begum;
my siblings, Habib, Waheed, Tahera;
my wife, Faiza;
and my son, Musaab.

Masood Ur Rehman

I dedicate this effort to my parents, Jamshed Iqbal and Nuzhut Jamshed;
my siblings, Laiba, Maliha, Mariam;
and my wife, Aqsa Tariq.

Muhammad Ali Jamshed

Contents

Editor Biography

Masood Ur Rehman received a B.Sc. degree in electronics and telecommunication engineering from the University of Engineering and Technology, Lahore, Pakistan, in 2004 and a M.Sc. and Ph.D. degrees in electronic engineering from Queen Mary University of London, London, UK, in 2006 and 2010, respectively. He worked at Queen Mary University of London as a Postdoctoral Research Assistant until 2012 before joining the Centre for Wireless Research at the University of Bedfordshire as a Lecturer. He served briefly at the University of Essex, UK and then moved to the James Watt School of Engineering at the University of Glasgow, UK in the capacity of an Assistant Professor. His research interests include compact antenna design, radiowave propagation and channel characterization, satellite navigation system antennas in cluttered environment, electromagnetic wave interaction with human body, body-centric wireless networks and sensors, remote health care technology, mmWave and nano-communications for body-centric networks, and D2D/H2H communications. He has worked on a number of projects supported by industrial partners and research councils. He has contributed to a patent and authored/co-authored 4 books, 7 book chapters, and more than 120 technical articles in leading journals and peer reviewed conferences. Dr. Ur Rehman is a fellow of the Higher Education Academy, UK, a member of the IET and part of the technical program committees and organizing committees of several international conferences, workshops, and special sessions. He is acting as an Associate Editor of the IEEE Access and IET Electronics Letters and Lead Guest Editor of numerous special issues of renowned journals. He also serves as a reviewer for book publishers, IEEE conferences, and leading journals.

Muhammad Ali Jamshed received a Ph.D. degree from the University of Surrey, Guildford, UK, in 2021. He is endorsed by Royal Academy of Engineering under exceptional talent category. He was nominated for Departmental Prize for Excellence in Research in 2019 at the University of Surrey. He served briefly as Wireless Research Engineer at BriteYellow Ltd., UK, and then moved to James Watt School of Engineering, University of Glasgow, as a Post-Doctoral Research Assistant. He has authored/co-authored 2 book chapters and more than 37 technical articles in leading journals and peer reviewed conferences. His main research interests include EMF exposure reduction, low SAR antennas for mobile handsets, machine learning for wireless communication, Backscatter communication, and wireless sensor networks. He served as a Reviewer, TPC, and the Session Chair, at many well-known conferences, i.e. ICC, WCNC, VTC, GlobeCom etc., and other scientific workshops.

List of Contributors

Yasir Alfadhl
School of Electronic Engineering and
Computer Science
Queen Mary University of London
London
UK

Tim W.C. Brown
Institute of Communication Systems
(ICS)
Home of 5G and 6G Innovation
Centre, University of Surrey
Guildford
UK

Xiaodong Chen
School of Electronic Engineering and
Computer Science
Queen Mary University of London
London
UK

Fabien Héliot
Institute of Communication Systems
(ICS)
Home of 5G and 6G Innovation
Centre, University of Surrey
Guildford
UK

Muhammad Ali Imran
James Watt School of Engineering
University of Glasgow
Glasgow
UK

Muhammad Ali Jamshed
James Watt School of Engineering
University of Glasgow
Glasgow
UK

Wali Ullah Khan
Interdisciplinary Center for Security
Reliability and Trust (SnT)
University of Luxembourg
Luxembourg City
Luxembourg

Sung Won Kim
Department of Information and
Communication Engineering
Yeungnam University
Gyeongsan-si
South Korea

Ali Nauman
Department of Information and
Communication Engineering
Yeungnam University
Gyeongsan-si
South Korea

Haris Pervaiz
School of Computing and
Communications
Lancaster University
Lancaster
UK

Muhammad Rafaqat Ali Qureshi
School of Electronic Engineering and
Computer Science
Queen Mary University of London
London
UK

Masood Ur Rehman
James Watt School of Engineering
University of Glasgow
Glasgow
UK

Preface

The past decade has seen a huge upsurge in the demand of wireless devices that are expected to cross the 29.4 billion mark by 2030. This increase is fueled by the advances in wearables, portables, flexible electronics, and other wireless technologies facilitating communication, transportation, and navigation needs of billions of users around the world in the wake of Internet of Things and 5G/6G. These rising numbers, along with ever-growing data requirements, necessitate a growth in the capacity of wireless communication networks by almost 1000 times. Part of this capacity enhancement will be made possible by increasing the number of access points (APs). These developments are ultimately resulting in added electromagnetic field (EMF) exposure sources in the environment.

EMF exposure has been deemed prone to inflict adverse health and safety effects on the users. The World Health Organization (WHO) has classified these EMF radiations as possibly carcinogenic to humans and has an ongoing project to assess potential health effects of exposure to EMF in the general and working population. The Federal Communications Commission (FCC) and the International Commission on Non-Ionizing Radiation Protection (ICNIRP) have, therefore, imposed strict safety standards for device operation. Consequently, EMF exposure characterization warranting strict adherence to these safety regulations is a vital design parameter for wireless devices to ensure the safety of the users.

The current developments and expected future growth of the wireless systems are also mounting concerns regarding users' safety and possible health consequences of EMF exposure to modern wireless technologies, such as millimeter-wave (mmWave) communications, massive multiple-input multiple-output (MIMO), and beamforming. It necessitates deeper investigations on health risk assessments and requires a comprehensive reference dealing with this fundamental and paramount issue suggesting some novel directions for updating the EMF exposure evaluation framework.

A dedicated book tackling this important issue is seldom available. Therefore, this volume will not only fill this gap but also educate the reader on most important aspects of designing energy efficient and low EMF wireless devices laying foundation for future advancements. A multidisciplinary approach based on artificial intelligence (AI) and new multiplexing technologies like non-orthogonal multiple access (NOMA) is adopted to devise efficient mechanisms and techniques realizing low EMF solutions through integration of antenna design, system modeling, and signal processing.

Both software and hardware solutions to minimize EMF exposure covering state-of-the-art and advanced topics are discussed. EMF evaluation tools and numerical assessment methods for conventional as well as future wireless systems' enablers such as mmWave technologies are detailed as also is EMF reduction through radio resource allocation, energy conservation, EMF-aware antenna design, backscatter communication, and MIMO. Moreover, a comprehensive account of validation studies as well as the modeling and selection of dielectric properties for all the age groups are utilized to provide sufficient background and highlight recent advancements. The book is concluded by highlighting potential future directions of research and implementation for energy-efficient and low EMF user proximity wireless devices. The book covers a wide variety of subject categories and would, therefore, benefit a larger readership in the scientific community.

University of Glasgow
Glasgow, UK

Masood Ur Rehman
Muhammad Ali Jamshed

1

Electromagnetic Field Exposure: Fundamentals and Key Practices

Muhammad Ali Jamshed[1], Fabien Héliot[2], Tim W.C. Brown[2], and Masood Ur Rehman[1]

[1]*James Watt School of Engineering, University of Glasgow, Glasgow, UK*
[2]*Institute of Communication Systems (ICS), Home of 5G and 6G Innovation Centre, University of Surrey, Guildford, UK*

1.1 Introduction

In the past, significant research efforts have been devoted to first understanding how EM field (EMF) exposure affects humans [1–3] and, then, to create tools for measuring exposure and defining exposure metrics [4–6]; these measuring techniques and metrics can be used to establish exposure recommendations [7]. Indeed, the health impact of EMF, magnetic field (MF), and electrical field (EF) is currently being contested in studies and among the general public, particularly for children [8]. Wireless communication (e.g. the cellular system) has grown so rapidly in recent decades that it is now one of the most major sources of EMF exposure in the general environment (see Figure 1.1). Similarly to other sources of EMF exposure, measures and recommendations have been created in wireless communication throughout the last decades [10] to restrict exposure and, thereafter, enhance approaches to minimize it [11]. In the future generation of communication networks, the exponential increases in both multimedia traffic and connected devices will necessitate a rise in the number of access points (APs) (e.g. base stations) to meet demand. As a result of the rising number of wireless devices and APs, the level of EMF exposure will increase. Similarly, the widespread use of mmWave spectrum in 5G, which will have carrier frequencies over 24 GHz, is anticipated to have an effect on exposure since it would necessitate a high density of APs [12]. Recent research in [13–16] has revealed that exposure at these frequencies may pose some health risks.

Figure 1.1 Common EMF exposure sources generally present in the environment [9].

The chapter's structure and key topics of discussion are summarized as follows:

1. Section 1.2 covers the existing techniques for assessing EMF exposure in various circumstances, i.e. the EMF assessment framework, and includes information on the metrics most typically used for measuring EMF exposure in communication. First, research projects relating to the EMF exposure assessment frameworks are provided; the majority of these studies outline their EMF exposure evaluation mechanism, examine the reasons of exposure, and then recommend solutions to minimize it. Second, different categories of exposure metrics are reviewed, where each category of metrics is explained vis–á–vis its target scenario(s). Third, generic metrics are presented, which are developed by integrating measurements from several categories.

2. Section 1.3 explains and illustrates how the various available EMF metrics have been utilized for restricting (i.e. creating standards) or lowering exposure.
3. Finally, Section 1.4 concludes the chapter.

1.2 EMF Metric and Evaluation Framework

A significant amount of work has been carried out in recent years for evaluating the EMF exposure in various scenarios, using different measurement systems and tools, to assess the potential risks emanating from EMF radiations in wireless communications and mitigate their effects (through guidelines and EMF-aware reduction techniques). As a result, EMF monitoring has gained relevance in wireless networks over the last decade [17], given that ambient RF-EMF exposure does not remain constant over time owing to environmental changes and variations in the number of active users (as well as the nature of their device usage). For example, the moniT (acronym for electromagnetic radiation exposure assessment in mobile communications) project, funded by Optimus, TMN, and Vodafone [18], provided public information on population exposure to EMF from mobile communication systems in Portugal from 2004 to 2012. This project's monitoring system was built on a network of autonomous remote probing stations and a comprehensive EMF sounding program, both of which were carried out in public spaces around the country. According to the project monitoring data, the EMF values of mobile systems were below the required threshold. Another EMF assessment and monitoring effort was the SEMONT project, which was implemented and utilized for real-time EMF exposure evaluation. Monitoring findings indicated that possible exposure was well below the permissible level set by Serbian legislation for the general population [19]. Their approach was then utilized to quantify the exposure produced by GSM when fluctuations in traffic circumstances were considered [20]. According to RF exposure assessments, exposure levels tend to grow with rising urbanization [21]. Meanwhile, the exposure survey assessment in [22] discovered that exposure levels in Europe are not exceeding the recommended levels, but exposure from wireless communication devices has increased significantly over the last years, accounting for more than 60% of total exposure.

In addition to these monitoring initiatives, other projects, such as the monitoring and control activities relating to electromagnetic fields in the RF range (MONICEM) and low EMF exposure future network (LEXNET) projects, have established new EMF assessment metrics that may be used to reduce the overall level of EMF exposure. For example, in MONICEM, which was supported by both the inter-University center for the study of interactions between electromagnetic fields and biosystems (ICEmB) and the institute for environmental protection and research (ISPRA), it was discovered that services such as cellular base stations, wireless networks, and so on create large amounts of EMF radiations, much over the natural limitations. The project created an environmental impact indicator (FIAE) based on the EMF derived from a generic source [23]. Similarly, in the LEXNET project, which was funded by the European Commission, a new realistic metric known as the exposure index (EI) [24] was developed to quantify the degree of EMF exposure to people in the environment. Using this criterion, the research established innovative strategies for lowering (by at least 50%) human exposure to electromagnetic (EM) radiation generated by wireless communication while maintaining quality of service (QoS) [25]. The metrics created in MONICEM or LEXNET are intended for assessing or realistically modeling EMF exposure across vast geographical regions while accounting for various forms of EMF radiations. These more general measurements or assessment frameworks sometimes rely on or combine existing measures created for more particular contexts. For example, consider the EI created by LEXNET, which includes in its definition the specific absorption rate (SAR) and power density (PD), both of which are typical metrics for measuring the EMF exposure of wireless communication devices and equipment. In the following sections, we will first go through the most often used metrics in wireless communications for analyzing EMF exposure in various circumstances, and then describe how some of them may be combined to generate more general metrics.

1.2.1 EMF Exposure Factors

As a byproduct of its transmission, each device delivering information to another device creates EMF exposure to users or persons in wireless communications. In general, the total exposure at any location in a given region under observation is the sum of the radiations emitted by all transmitting devices in the vicinity (accounting for both the active and passive exposures). The severity of the exposure is determined by four major criteria, which are discussed in the following.

1.2.1.1 Transmit Antenna Regions
Transmitting antennas typically have two radiating regions: near field and far field, with the near field region further classified as reactive and radiated near field dependent on the distance and frequency of the radiating antenna. The reactive near field lies in the immediate proximity of the antenna, where the electric and MF are 90° out of phase, making the reactive effect more dominating. The radiating near field, also known as the Fresnel area, is the space between the reactive near and far fields. In this area, the radiating impact of the antenna begins to outweigh the reactive effect. The far field area, on the other hand, is further away from the antenna and has the electric and MF in phase. It should be noted that each zone is determined by specific boundary criteria, which are further specified

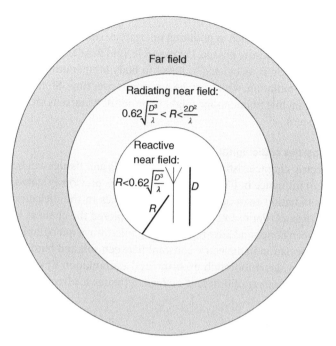

Figure 1.2 Antenna field areas are depicted [26].

in Figure 1.2. In Figure 1.2, D denotes the diameter of the antenna, R the radius of each zone, and λ the wavelength of an EM wave. The impact of the near field on EMF exposure is more significant in the uplink scenario, when the antenna(s) of a user mobile device radiate(s) to send data to an access point (AP) and most of the antenna(s) dissipated energy can easily be absorbed by the user body/head (given the user body/presence head's in the near field region) [27]. The influence of the radiated EMF, on the other hand, decreases with distance in the far field. It should be emphasized that active exposure normally results from near field EMF waves, whereas passive exposure typically results from far field radiations.

1.2.1.2 Transmit Antenna Characteristics

The transmitting antenna's parameters, such as transmitted power, antenna gain, directivity, effective aperture, polarization, beam width, and so on, are critical in defining the extent of exposure. The intensity of exposure is generally proportional to the intensity of the EF, which is proportional to the transmit power. For example, in [28], the EMF radiations from mobile communication antennas were examined by taking into consideration the relevance of antenna characteristics for determining exposure.

1.2.1.3 Duration of Exposure

As with any other sort of exposure, such as pollution or cigarette smoke, the longer the exposure, the greater the exposure dosage. For example, [29] has demonstrated that the duration of exposition is associated to a rise in body temperature when humans are exposed to RF radiation, which can be hazardous over time. Similarly, [30] claims that growing mobile phone usage might have negative impacts on the human reproductive system.

1.2.1.4 Electrical Properties of Biological Tissues

Variations in the dielectric characteristics of organic materials and tissues can be regarded as a significant influence in EMF exposure. Indeed, as previously stated, children absorb more radiation than adults due to differences in the dielectric characteristics of their tissues. For example, [31], which explored the changes in dielectric constant between bones and fatty regions using microwave tomography, found a relatively large deviation in dielectric constant between soft and hard tissues. Meanwhile, [32] provides a thorough experimental examination linked to the variation in dielectric constant of different biological tissues for frequencies ranging from 10 Hz to 20 GHz.

1.2.2 EMF Exposure Metrics

Several metrics have been defined throughout the years in order to analyze and predict the EMF exposure of wireless communication systems in various circumstances, depending on the numerous parameters indicated in Section 1.2.1. To the best of our knowledge, there are four primary categories of EMF metrics, namely, SAR, PD, exposure-ratio, and dosage, which may be grouped as shown in Figure 1.3.

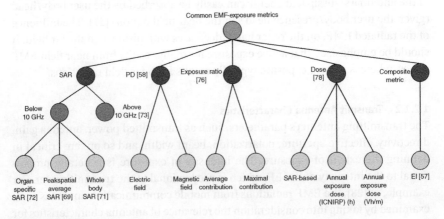

Figure 1.3 The most often used metrics for assessing EMF exposure [9].

1.2.2.1 Specific Absorption Rate

The SAR is a measure of the generated EMF inside the human body when exposed to a transmitting antenna's near field. Watts per kilogram are the units of measurement of SAR. The SAR measure is widely used by regulatory organizations throughout the world to determine exposure standards and evaluate the exposure produced by various handset [35]. Indeed, to ensure public safety, each handset maker should give the electromagnetic energy deposition within surrounding biological tissues, as measured by the SAR [36, 37]. The SAR in the near field of an antenna mounted on a wireless device can be expressed as [38];

$$\text{SAR} = \frac{\sigma \times E^2}{m_d} \text{ (W/kg)}. \tag{1.1}$$

In (1.1), σ represents the conductivity of the exposed tissue(s), E indicates the strength of the EF and m_d is the mass density of the sample under test. Figure 1.4 depicts a typical setup for measuring the SAR of a human head, in which a radio frequency (RF) radiating device (with two antenna components in our example) is positioned close to a phantom head, and a probe (receiver) is used to measure the strength of E [33]. To test SAR in the worst case scenario, the phantom head would be filled with a sugar solution that replicated the dielectric and conduction characteristics of brain tissue on average. The SAR may be further classified

Figure 1.4 A typical SAR measuring setup is depicted. Source: Jamshed et al. [9]/with permission of IEEE.

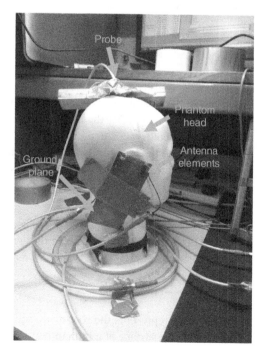

based on the EMF absorbed by different areas of the human body as whole body averaged SAR, organ-specific SAR, and peak spatial average SAR [39, 40].

In confined contexts (i.e. rooms), the whole-body averaged SAR or global SAR may be measured by measuring the reverberation time with and without humans within the room; the whole-body SAR is then approximated based on the difference in reverberation time [41]. The organ-specific SAR or local SAR is used to estimate the radiation absorption of a given organ inside the human body, and it is averaged spatially over the mass of a certain organ or tissue in the body [42]. Local SAR medicinal consequences are localized to a single bodily tissue averaged over 1 g or 10 g. In contrast to local SAR, global SAR considers the biological impacts on the entire body. In conjunction with the preceding SAR definitions, the peak-spatial average SAR is used to determine the limits of SAR absorption for different areas of the human body, as well as to offer guidelines for safeguarding humans from RF near field exposure [40]. Meanwhile, for frequencies over 24 GHz, the energy contribution received by biological tissues is quite minimal in the reactive near field. Indeed, the average SAR becomes null for frequencies higher than 10 GHz due to the shallow penetration depth [43]; thus, the point-wise power loss density (PLD) methodology is typically used to estimate the correct radiations absorbed by the human body and obtain accurate exposure measurement in the mmWave frequency band. The following equation illustrates the relationship between PLD and SAR [43];

$$\text{PLD}_{pnt} = \sigma \times |E^2| = \rho \times \text{SAR}_{pnt} \ (\text{W/kg}), \tag{1.2}$$

where ρ defines the mass density of the sample under test.

1.2.2.2 Power Density

In contrast to the SAR, which is beneficial for assessing EMF exposure in the near field of an antenna, the PD is the metric of choice for measuring EMF exposure in the far radiating field of an antenna, and is measured in Watts per square meter. In general, [26] gives the PD of an isotropic antenna in its far field, which is uniform (power per unit area) in all directions, and is as follows;

$$\text{PD} = \frac{P_t}{4\pi R^2} (\text{W/m}^2), \tag{1.3}$$

where P_t is the transmitting power of the transmit antenna and R is the distance at which the PD is measured. Whereas in the context of human body exposure, the PD of a transmitting antenna in its far field region can be defined as [44]

$$\text{PD} = \frac{|E_i^2|}{\eta} (\text{W/m}^2), \tag{1.4}$$

where E_i (V/m) represents the root-mean-squared value of the EF strength incident on the tissue surface of a human body and η (V/A) is the wave impedance.

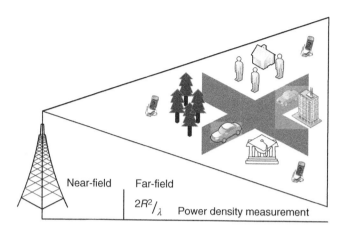

Figure 1.5 An overview of PD measurement. Source: Jamshed et al. [9]/with permission of IEEE.

Furthermore, because EF and MF are in phase in this region, MF strength may be utilized to assess PD in the distant field. Figure 1.5 depicts an overview of PD measuring.

1.2.2.3 Exposure-Ratio
When there are several exposure sources, the exposure-ratio metric is used to calculate each source's contribution to overall exposure. It may be defined as the average or maximum contribution of several sources to the overall exposure value, such that

$$\text{Exposure-ratio} = v \left(\frac{S_{signal}}{S_{total}} \right) \times 100\%. \tag{1.5}$$

In (1.5), S_{signal} represents the PD of the RF signal at a geographical location v and S_{total} is the total power density of all the signals at the same location v. Because of its nature, this metric is very useful for determining the contributions of different radio access technologies (RATs) to overall exposure in cellular systems. The exposure levels in near and far fields were computed in [45] using (1.5), and it was established that the exposure level decreases with distance from the base station. Furthermore, it was demonstrated in the [46], using the exposure-ratio metric, that LTE (in comparison with the contributions from other RATs) only accounts for 4% of overall exposure in Stockholm. Meanwhile, in the [47], the notion of exposure-ratio is utilized to quantify RF exposure in various situations, and it was discovered that the average contribution to total exposure is more than 60% for GSM, more than 3% for UMTS-HSPA, and less than 1% for both LTE and WiMAX.

1.2.2.4 Dose

It should be noted that the SAR and PD are affected by the majority of the elements given in Section 1.2.1, with the exception of the exposure duration, i.e. time; this is the key difference between these metrics and the dose measure. The dose includes the total radiations absorbed by the sample under test over a given time period. The dose measurement is often referred to as the total SAR or PD over time. From SAR and PD, i.e. Equations (1.1) and (1.3), the dose metric can be defined as

$$\text{Dose} = \int \text{SAR} \times t \, dt \quad \text{[J/kg]}, \tag{1.6}$$

where t represents the total duration in which the sample under test stays in the near field of the radiating antenna.

1.2.2.5 Composite/Generic Metric of EMF Exposure

As previously stated, exposure metrics designed to measure a specific exposure, such as exposure produced by a specific device (SAR) or exposure at a specific geographical point (PD), can be combined to measure/model the exposure of a group of devices and/or large geographical areas. The EI established in [24] is a nice example of such composite EMF exposure metric. This metric incorporates both active (through SAR) and passive (via PD) exposures over time for various types of populations, settings, mobile devices, cellular equipment, usages, postures, and so on. It is defined in [24].

1.3 Application of Metric for Setting Guidelines/Limits and Reducing Exposure

Limits and standards have been set to safeguard the public from the recognized consequences of EMF exposure on health. For example, in the United Kingdom, the national radiological protection board (NRPB) has published guidance on the limits of EMF exposure from mobile phones, APs, and other sources in different publications [48, 49] since the mid-90s. Manufacturers and operators utilize these recommendations to conduct EMF compliance evaluations and verify that cellular equipment meets applicable regulatory standards for human exposure before it is placed on the market and installed on site. In Europe, the European Council approved a recommendation (1999/519/CE) on limiting EMF exposure to the general population in 1999. On partial body exposure, the Federal Communications Commission (FCC), in the United States, imposes an SAR (i.e. local SAR) threshold of 1.6 W/kg for 1 g of tissue and 2.0 W/kg for 10 g of tissue [33]. Meanwhile, the International Commission on Non-Ionizing Radiation Protection (ICNIRP) defines baseline limits for both active and passive EMF exposure, often in terms of SAR or PD. For example, the ICNIRP recently recommended that the entire body average SAR (i.e. global SAR) should not exceed 0.4 W/kg [34]. Furthermore, the exposure dose (based on SAR) should be restricted to 500 J/kg for less than 1 second [34]. More recently, it has been advocated in [50] that some of these limitations should be revised to account for the possible long-term effects of EMF exposure, given that most current limitations/guidelines are based on exposure metrics, such as SAR measurement in mobile handsets, that do not account for the duration of the exposure.

In recent years, there has been a significant amount of research community input linked to the creation of effective strategies for lowering EMF exposure in different generations of mobile communication systems, for example, [51–53]. In the next section, we provide some of these strategies based on the exposure metrics defined in Section 1.2.2 on which they rely to limit EMF exposure.

1.3.1 SAR Reduction

Given that the energy lost within the reactive near field zone is easily absorbed by the human body [27], EMF exposure is more intense in the near field. As a result, if the exposure is greater, there is more chance to reduce it. Given that the SAR is quite a popular metric for evaluating the EMF exposure, a plethora of works on exposure reduction have relied on this metric. In [54], it was discovered that decreased SAR not only minimizes EMF exposure but also boosts the efficiency of multi-antenna systems. The region of a specific head and handset configuration that exhibits the highest degree of SAR is generally connected with a region near the surface of the head where the handset is positioned. It is demonstrated by the analysis that the EF and MF in this region can be modified by using smart antennas so that the SAR can be reduced [54]. Similarly, a signal processing approach based on multi-antennas, referred to as SAR codes, is proposed in [55]; it is based on the idea that a portable device placed in a "talk position," i.e. placed close to the user's head, can reduce its user-imparted power by setting a phase offset of $\pi/3$ radians (or 60°) between the transmit spatial samples. This works first model the SAR as function of an offset phase, such that

$$\text{SAR}(\theta) = P(s_1 + s_2 \cos(\theta + \theta°)), \tag{1.7}$$

and then uses this equation to find the phase that minimizes the SAR. In (1.7), P represents the total power supplied by the user equipment to transmit data, s_1 and s_2 are positive real parameters that can be obtained based on measurements, θ is a relative phase, and $\theta°$ is the phase offset. It is shown in [55] that these SAR codes can be used to reduce the exposure towards human, and can extend the range of portable devices while maintaining a safe exposure level. Moreover, it is shown in [56] that the SAR codes can also be used to simultaneously maximize the far field performance while controlling near field SAR. Another recent antenna study shown that adding a conductive composite filled with carbon-silica hybrid filler may significantly lower the maximum SAR value (by up to 70%) [57]. SAR has also been used to reduce exposure at the system level; in [51], a power adaptation mechanism has been presented to reduce exposure in the heterogeneous networks (HetNets) uplink.

1.3.2 PD Reduction

The PD metric is used to quantify far field exposure; hence it is mostly used to assess passive or downlink exposure. In the downlink, the degree of exposure is judged safe for humans because the transmit power falls at least quadratically with distance as long as the human body is far enough away from any APs (e.g. if it resides outside the exclusion zone [58] of a macro base station). One easy technique to lower PD is to reduce the transmit power of APs; this is readily accomplished by transitioning to small-cell technology. Indeed, [52] discovered that employing femto-base stations instead of macro-base stations can lower PD by a factor of 20–40.

1.3.3 Exposure-Ratio Reduction

By decreasing the average and maximal contributions from each source, the exposure-ratio may be lowered. To minimize the total exposure-ratio, the SAR and PD exposure reduction approaches must be used together. To minimize the exposure-ratio, the approach for SAR reduction provided in [55] can be used with the technique for PD reduction proposed in [52].

1.3.4 Dose Reduction

Because exposure dosage is proportional to exposure duration, it can be reduced by limiting the use of wireless devices or lowering the amount of energy (power radiated over time) emitted towards the user. In this context, the studies in [59] and [60] used multi-user scheduling to significantly reduce (by up to two orders of magnitude) the active exposure dosage, in the uplink of single and multi-cell cellular systems, respectively. This is accomplished by utilizing the time dimension in a traditional orthogonal frequency division multiplexing (OFDM)-based resource allocation architecture, where the instantaneous transmit power of users is dispersed and reduced across numerous time intervals. It can also be lowered by enhancing transmission quality. For example, [61] showed that by connecting mobile users to a specialized small cell in a train setting while utilizing GSM technology at 1800 MHz, the exposure dosage of the brain and whole-body may be substantially lowered by a factor 35 and 11, respectively.

1.3.5 Composite EMF Exposure Reduction

Methods designed to limit the overall exposure of vast and complicated communication networks are generally the most successful, because they strive to reduce total exposure rather than focusing on a single source. For example, in [62], an EMF-Aware cell selection technique based on the EI has been suggested to limit the EMF exposure of HetNets systems while retaining QoS. This approach distributes the load among different cells in a HetNet system while also lowering the total EMF exposure. Another EI-based cell selection technique in the context of the HetNet system was proposed in [63]; it was discovered that progressive deployment of small cells can lower EI at the expense of QoS. Meanwhile, [53]has achieved a 50% reduction in EI for active users by creating a self-optimizing self-organizing improved inter-cell interference coordination approach. In the context of a wireless local area network (WLAN), an algorithm has been published in [64] to optimize the wireless configuration for minimizing the EI of a studied population; this technique achieves an EI reduction of up to 86%.

1.4 Conclusion

This chapter provides a comprehensive assessment of exposure evaluation, limiting, and mitigation for existing and future wireless devices and equipment. When it comes to assessing EMF exposure, this chapter has covered the most frequent assessment frameworks and metrics used in wireless communications to assess exposure. It also identifies the metrics, such as SAR and dosage, that are most suited for minimizing EMF exposure. This chapter also evaluated current exposure standards and described how they might be revised to better represent the real nature of EMF exposure, i.e. by taking exposure duration into consideration.

References

1 Warren, S. and Whipple, G. (1923). Roentgen ray intoxication: I. Bacterial invasion of the blood stream as influenced by X-ray destruction of the mucosal epithelium of the small intestine. *Journal of Experimental Medicine* 38 (6): 713–723.

2 Taylor, H.D., Witherbee, W.D., and Murphy, J.B. (1919). Studies on X-ray effects: I. Destructive action on blood cells. *The Journal of Experimental Medicine* 29 (1): 53–73.

3 Aldrich, T.E. and Easterly, C.E. (1987). Electromagnetic fields and public health. *Environmental Health Perspectives* 75: 159–171.

4 Baris, D., Linet, M.S., Tarone, R.E. et al. (1999). Residential exposure to magnetic fields: an empirical examination of alternative measurement strategies. *Occupational and Environmental Medicine* 56 (8): 562–566.

5 Kleinerman, R.A., Linet, M.S., Hatch, E.E. et al. (1997). Magnetic field exposure assessment in a case-control study of childhood leukemia. *Epidemiology* 8 (5): 575–583.

6 Russell, J. (1986). Assessment of priorities when introducing some radiation protection methods in radio diagnosis. *The British Journal of Radiology* 59 (698): 153–156.

7 Martin, A. (1990). Health effects of electromagnetic fields. *Journal of the Royal Society of Medicine* 83 (10): 670–672.

8 Otto, M. and von Mühlendahl, K.E. (2007). Electromagnetic fields (EMF): do they play a role in children's environmental health (CEH)? *International Journal of Hygiene and Environmental Health* 210 (5): 635–644.

9 Jamshed, M.A., Heliot, F., and Brown, T.W. (2019). A survey on electromagnetic risk assessment and evaluation mechanism for future wireless communication systems. *IEEE Journal of Electromagnetics, RF and Microwaves in Medicine and Biology* 4 (1): 24–36.

10 National Research Council et al. (1993). *Assessment of the Possible Health Effects of Ground Wave Emergency Network*. National Academies Press.

11 Popović, M., Tesanovic, M., and Radier, B. (2014). Strategies for reducing the global EMF exposure: cellular operators perspective. *2014 11th International Symposium on Wireless Communications Systems (ISWCS)*, pp. 836–841, IEEE.

12 Busari, S.A., Huq, K.M.S., Mumtaz, S. et al. (2017). Millimeter-wave massive MIMO communication for future wireless systems: a survey. *IEEE Communication Surveys and Tutorials* 20 (2): 836–869.

13 Le Dréan, Y., Mahamoud, Y.S., Le Page, Y. et al. (2013). State of knowledge on biological effects at 40–60 GHz. *Comptes Rendus Physique* 14 (5): 402–411.

14 Debouzy, J., Crouzier, D., Dabouis, V. et al. (2007). Biologic effects of millimeteric waves (94 GHz). Are there long term consequences? *Pathologie-Biologie* 55 (5): 246–255.

15 Walters, T.J., Blick, D.W., Johnson, L.R. et al. (2000). Heating and pain sensation produced in human skin by millimeter waves: comparison to a simple thermal model. *Health Physics* 78 (3): 259–267.

16 Di Ciaula, A. (2018). Towards 5G communication systems: are there health implications? *International Journal of Hygiene and Environmental Health* 221 (3): 367–375.

17 Joseph, W., Verloock, L., Tanghe, E., and Martens, L. (2009). In-situ measurement procedures for temporal RF electromagnetic field exposure of the general public. *Health Physics* 96 (5): 529–542.

18 Oliveira, C., Sabastiao, D., Carpinteiro, G. et al. (2007). The moniT project: electromagnetic radiation exposure assessment in mobile communications. *IEEE Antennas and Propagation Magazine* 49 (1): 44–53.

19 Djuric, N., Kljajic, D., Kasas-Lazetic, K., and Bajovic, V. (2015). The SEMONT continuous monitoring of daily EMF exposure in an open area environment. *Environmental Monitoring and Assessment* 187 (4): 1–17.

20 Pasquino, N. (2017). Measurement and analysis of human exposure to electromagnetic fields in the GSM band. *Measurement* 109: 373–383.

21 Sagar, S., Adem, S.M., Struchen, B. et al. (2018). Comparison of radiofrequency electromagnetic field exposure levels in different everyday microenvironments in an international context. *Environment International* 114: 297–306.

22 Gajšek, P., Ravazzani, P., Wiart, J. et al. (2015). Electromagnetic field exposure assessment in Europe radiofrequency fields (10 MHz–6 GHz). *Journal of Exposure Science and Environmental Epidemiology* 25 (1): 37–44.

23 ISPRA (2011). MONICEM: Monitoring and control activities relating to electromagnetic fields in the radio frequency (RF) range. http://www.isprambiente .gov.it/en/publications/reports/monicem-monitoring-and-control-activities-relating (accessed 11 November 2018).

24 Varsier, N., Plets, D., Corre, Y. et al. (2015). A novel method to assess human population exposure induced by a wireless cellular network. *Bioelectromagnetics* 36 (6): 451–463.

25 Tesanović, M., Conil, E., De Domenico, A. et al. (2014). Wireless networks and EMF-paving the way for low-EMF networks of the future: the LEXNET project. *IEEE Vehicular Technology Magazine* 9 (2): 20–28.

26 Balanis, C.A. (1997). *Antenna Theory: Analysis and Design*. Wiley.

27 Swerdlow, A. et al. (2012). Health Effects from Radio Frequency Electromagnetic Fields. *Report of AGNIR*. Health Protection Agency.

28 Baltrėnas, P., Buckus, R., and Vasarevičius, S. (2012). Research and evaluation of the intensity parameters of electromagnetic fields produced by mobile communication antennas. *Journal of Environmental Engineering and Landscape Management* 20 (4): 273–284.

29 Shellock, F.G., Schaefer, D.J., and Crues, J.V. (1989). Alterations in body and skin temperatures caused by magnetic resonance imaging: is the recommended exposure for radio frequency radiation too conservative? *The British Journal of Radiology* 62 (742): 904–909.

30 Gye, M.C. and Park, C.J. (2012). Effect of electromagnetic field exposure on the reproductive system. *Clinical and Experimental Reproductive Medicine* 39 (1): 1–9.

31 Morris, R.D., Morgan, L.L., and Davis, D. (2015). Children absorb higher doses of radio frequency electromagnetic radiation from mobile phones than adults. *IEEE Access* 3: 2379–2387.

32 Gabriel, S., Lau, R., and Gabriel, C. (1996). The dielectric properties of biological tissues: II. Measurements in the frequency range 10 Hz to 20 GHz. *Physics in Medicine & Biology* 41 (11): 2251–2269.

33 Radiofrequency Electromagnetic Fields (1997). Evaluating compliance with FCC guidelines for human exposure to radio frequency electromagnetic fields. *OET Bull* 65: 1–53.

34 ICNIRP (2018). Guidelines for limiting exposure to time-varying electric, magnetic and electromagnetic fields (100 kHz to 300 GHz). https://www.icnirp.org/cms/upload/consultation-upload/ICNIRP-RF-Guidelines-PCD-2018-07-11.pdf (accessed 26 February 2019).

35 Christian (2018). Cell phone radiation charts (SAR) levels of popular phones. https://emfacademy.com/cell-phone-radiation-charts-sar-levels-popular-phones/ (accessed 15 November 2018).

36 Chen, J.-Y., Gandhi, O.P., and Conover, D.L. (1991). SAR and induced current distributions for operator exposure to RF dielectric sealers. *IEEE Transactions on Electromagnetic Compatibility* 33 (3): 252–261.

37 Beard, B.B., Kainz, W., Onishi, T. et al. (2006). Comparisons of computed mobile phone induced SAR in the SAM phantom to that in anatomically correct models of the human head. *IEEE Transactions on Electromagnetic Compatibility* 48 (2): 397–407.

38 Lin, J.C. (2000). Specific absorption rates (SARs) induced in head tissues by microwave radiation from cell phones. *IEEE Antennas and Propagation Magazine* 42 (5): 138–139.

39 IEEE (1992). *IEEE Standard for Safety Levels with Respect to Human Exposure to Radio Frequency Electromagnetic Fields, 3kHz to 300 GHz*. Institute of Electrical and Electronics Engineers, Incorporated.

40 Ahlbom, A., Bergqvist, U., Bernhardt, J. et al. (1998). Guidelines for limiting exposure to time-varying electric, magnetic, and electromagnetic fields (up to 300 GHz). *Health Physics* 74 (4): 494–521.

41 Bamba, A., Joseph, W., Andersen, J.B. et al. (2012). Experimental assessment of specific absorption rate using room electromagnetics. *IEEE Transactions on Electromagnetic Compatibility* 54 (4): 747–757.

42 Thielens, A., Vermeeren, G., Joseph, W., and Martens, L. (2013). Stochastic method for determination of the organ-specific averaged SAR in realistic environments at 950 MHz. *Bioelectromagnetics* 34 (7): 549–562.

43 Li, J., Yan, S., Liu, Y. et al. (2017). A high-order model for fast estimation of electromagnetic absorption induced by multiple transmitters in portable devices. *IEEE Transactions on Antennas and Propagation* 65 (12): 6768–6778.

44 Nasim, I. and Kim, S. (2017). Human exposure to RF fields in 5G downlink. *arXiv preprint arXiv:1711.03683*.

45 Buckus, R., Strukčinskienė, B., Raistenskis, J. et al. (2017). A technical approach to the evaluation of radio frequency radiation emissions from mobile telephony base stations. *International Journal of Environmental Research and Public Health* 14 (3): 1–18.

46 Joseph, W., Verloock, L., Goeminne, F. et al. (2010). Assessment of general public exposure to LTE and RF sources present in an urban environment. *Bioelectromagnetics* 31 (7): 576–579.

47 Joseph, W., Verloock, L., Goeminne, F. et al. (2012). Assessment of RF exposures from emerging wireless communication technologies in different environments. *Health Physics* 102 (2): 161–172.

48 NRPB (National Radiological Protection Board) (1993). Restrictions on human exposure to static and time varying electromagnetic fields and radiation: scientific basis and recommendation for implementation of the board's statement. *Doc NRPB* 4 (5): 8–69.

49 Doll, S.R. (1992). *Electromagnetic Fields and the Risk of Cancer*. National Radiological Protection Board.

50 Hardell, L.O., Carlberg, M., Söderqvist, F. et al. (2007). Long-term use of cellular phones and brain tumours-increased risk associated with use for greater than 10 years. *Occupational and Environmental Medicine* 64 (9): 626–632.

51 Nasir, A., Shakir, M.Z., Qaraqe, K., and Serpedin, E. (2013). On the reduction in specific absorption rate using uplink power adaptation in heterogeneous small-cell networks. *2013 7th IEEE GCC Conference and Exhibition (GCC)*, pp. 474–478.

52 Aerts, S., Plets, D., Verloock, L. et al. (2013). Assessment and comparison of total RF-EMF exposure in femtocell and macrocell base station scenarios. *Radiation Protection Dosimetry* 162 (3): 236–243.

53 Sidi, H.B. and Altman, Z. (2016). Small cells' deployment strategy and self-optimization for EMF exposure reduction in HetNets. *IEEE Transactions on Vehicular Technology* 65 (9): 7184–7194.

54 Chim, K.-C., Chan, K.C., and Murch, R.D. (2004). Investigating the impact of smart antennas on SAR. *IEEE Transactions on Antennas and Propagation* 52 (5): 1370–1374.

55 Hochwald, B.M., Love, D.J., Yan, S., and Jin, J. (2013). SAR codes. *Information Theory and Applications Workshop (ITA), 2013*, pp. 1–9, IEEE.

56 Hochwald, B.M., Love, D.J., Yan, S. et al. (2014). Incorporating specific absorption rate constraints into wireless signal design. *IEEE Communications Magazine* 52 (9): 126–133.

57 Al-Ghamdi, A.A., Al-Hartomy, O.A., Al-Solamy, F.R. et al. (2017). Enhancing antenna performance and SAR reduction by a conductive composite loaded with carbon-silica hybrid filler. *AEU-International Journal of Electronics and Communications* 72: 184–191.

58 De Salles, A.A. and Fernández, C.R. (2006). Exclusion zones close to wireless communication transmitters aiming to reduce human health risks. *Electromagnetic Biology and Medicine* 25 (4): 339–347.

59 Sambo, Y.A., Héliot, F., and Imran, M. (2017). Electromagnetic emission-aware scheduling for the uplink of multicell OFDM wireless systems. *IEEE Transactions on Vehicular Technology* 66 (9): 8212–8222.

60 Sambo, Y., Héliot, F., and Imran, M. (2016). Electromagnetic emission-aware schedulers for the uplink of OFDM wireless communication systems. *IEEE Transactions on Vehicular Technology* 66 (2): 1313–1323.

61 Aerts, S., Plets, D., Thielens, A. et al. (2015). Impact of a small cell on the RF-EMF exposure in a train. *International Journal of Environmental Research and Public Health* 12 (3): 2639–2652.

62 De Domenico, A., Díez, L.F., Agüero, R. et al. (2015). EMF-Aware cell selection in heterogeneous cellular networks. *IEEE Communications Letters* 19 (2): 271–274.

63 Sidi, H.B., Altman, Z., and Tall, A. (2014). Self-optimizing mechanisms for EMF reduction in heterogeneous networks. *2014 12th International Symposium on Modeling and Optimization in Mobile, Ad Hoc, and Wireless Networks (WiOpt)*, pp. 341–348, IEEE.

64 Plets, D., Vermeeren, G., Poorter, E.D. et al. (2017). Experimental optimization of Exposure Index and quality of service in Wlan networks. *Radiation Protection Dosimetry* 175 (3): 394–405.

2

Exposure to Electromagnetic Fields Emitted from Wireless Devices: Mechanisms and Assessment Methods

Yasir Alfadhl

School of Electronic Engineering and Computer Science, Queen Mary University of London, London, UK

2.1 Fundamentals of EMF Interactions with the Human Body

The understanding of the interaction of electromagnetic fields (EMFs with living species) is a complicated subject due to the complex inhomogeneous nature of biological matter. In view of the difficulty of this topic, a wide number of studies have been conducted to determine various biological implications due to EMF exposure.

In the past few decades, numerous studies have contributed to the knowledge in the areas of interaction of EMFs with biological matter, as well as medical applications of non-ionizing radiation. Based on the current scientific literature, the World Health Organization (WHO) concluded that there is no evidence linking any health consequences from exposure to low level EMFs [1, 2]. However, some gaps in knowledge about the influence of EMFs on biological matter across the electromagnetic spectrum that needs further research.

In general, the human body is continually exposed to a complex mix of weak EM fields from a range of sources ranging from natural solar emissions to man-made sources such as power transmission lines and communication devices. With the continual evolution of EM applications, each individual is subjected to EM fields of different intensities and characteristics. Examples of such exposures can be demonstrated by considering a person exposed to the radiations from mobile phone signals (with various communication generation bands, Bluetooth, Wi-Fi, etc.); the same person may also be subjected to far-field radiations from stations, and emissions from devices of other uses, in addition to emissions

Low Electromagnetic Field Exposure Wireless Devices: Fundamentals and Recent Advances, First Edition.
Edited by Masood Ur Rehman and Muhammad Ali Jamshed.

from other broadcasting antennas. Humans may also experience exposure to EM fields while undergoing controlled medical treatments where the intensity of the field, frequency, and duration of exposure are monitored to achieve the desired biological effect.

Interactions of EMFs with the human body include various mechanisms depending on the nature of the source and the frequency portion of the electromagnetic spectrum. In addition to the frequency, the interactions are also influenced by the modulation scheme, power level, body orientation, electromagnetic properties of the body, and other factors relating to the exposure conditions. This chapter focuses on exposure to EMFs emitted from communication devices in the microwave and millimeter waves which constitute the fundamental frequencies used in communication devices.

Understanding the underlying interaction mechanisms caused by RF exposure is necessary for assessing the possible impact on human health. The interaction of EM fields with biological systems can be categorized by several mechanisms depending on the type and frequency of exposure.

The first logical step towards understanding this topic is the characterization of the EM field strengths and distributions within the exposed body or tissue. In practice, numerical dosimetry is applied as direct field measurement inside the living species using an invasive probe technique is very difficult if not impossible. In order to enable experimentations to assess the interaction of EMF with tissues inside the body (in vivo), various experimental techniques and numerical solutions can be employed. Other types of studies, however, are based on evaluating the fields inside tissue samples in isolation from the body (in vitro). Despite the apparent simplicity of assessment either in vivo or in vitro, many experimental and numerical difficulties are generated as a result of the inhomogeneity of the medium, relatively large size compared to the wavelength, smaller size or the thickness of the tissue sample, and so on. Furthermore, the utilization of thin probes to measure the fields may add to the complexity of the problem due to the introduction of a singular point at the tip of the probe where the field tends to intensify. By applying the correct numerical and computational dosimetry techniques, better understanding can be achieved for the underlying interaction mechanisms.

Previous studies, thus-far, have hypothesized different categories of the interaction of EM fields with human body and biological tissue, over a frequency band ranging from just above zero Hertz (Hz) to 300 GHz. These categories varied from induction of currents within tissues in the extremely low frequency (ELF) band, through the induction of thermal energy due to RF radiation absorption, all the way to the surface interactions at the mm-wave bands and beyond.

Exposure to non-continuous (pulsed) waves has a different impact on biological systems. Evaluation of such impacts is dependent on a variety of parameters, such as the intensity, pulse shape and width, and the pulse type (single or multiple). It is also worth that high-power pulses are widely utilized in medical applications, such as electroporation.

In general, the biological effects arising due to exposure to non-ionizing radiations are categorized in terms of thermal and non-thermal effects. Biological effects are expected to appear in response to induced heating from the EM power absorbed within the tissue or body. In addition, other effects may exist as a result of the direct interactions of E-fields or H-fields with the body cells or tissues.

2.1.1 Thermal Effect

The thermal effect constitutes a category of biological effects due to the temperature rise caused by the absorption of the EM energy within a given mass of biological material. Hypothetically, the basic definition of the electrical power flow is determined by the equation $P = I^2 R$, where I and R correspond to the total current and resistance of the material sample, respectively.

The EM power dissipated within a given mass can also be expressed in terms of the power absorbed per unit volume P_a, and can be defined as a function of material resistivity r and current density J:

$$P_a = r \cdot J^2 \ (W/m^3) \tag{2.1}$$

The power absorption can also be expressed in terms of the rms value of the electric field (E) and the material conductivity (σ), as shown as follows:

$$P_a = \sigma E^2 \ (W/m^3) \tag{2.2}$$

The rate at which the EM power is absorbed can be used to determine whether such power levels may initiate a biological thermal response.

Temperature in tissues exposed to EM fields continues to rise until the heat input is balanced by the rate at which it is removed. The temperature is generally dissipated from the material by means of various mechanisms, such as conduction with other tissue types, convection through blood perfusion. In addition, heat transfer also occurs, but less evidently, through radiations to the surroundings (e.g. for tissues in contact with the outer skin) [3].

It has been estimated that several minutes are typically needed for temperature equilibrium during exposure to RF radiations. In view of this slow response, the equilibrium temperature resulting from the pulsed fields of mobile telecommunications is essentially determined by the average of the power absorbed. For this reason, the specific absorption rate (SAR) dosimetric concept is introduced to determine the power absorbed per unit mass of tissue.

Practically, thermal variations due to exposure can be measured directly on the outer skin, or can be assessed using the thermal modelling, complemented with time- or frequency-domain EM computation techniques [4, 5].

The relationship between SAR and the resulting temperature rise is complex, and is dependent on many parameters. The most problematic feature of temperature calculations is modelling the effects of blood flow on the heat transfer and metabolic temperature rise.

The traditional continuum heat-sink model, developed by Pennes [6], was found to give remarkably accurate description of the heat transfer in biological tissues in many circumstances. Nevertheless, numerous modifications have been suggested since. For example, heat deposition within the head can be computed by coupling the SAR with a thermal model [7]. This model includes the convective effects of discrete blood vessels, whose anatomy was determined using magnetic resonance angiography of a healthy volunteer.

Many researchers have carried out various numerical modelling of the EM power dissipation and bio-heat transfer within models equivalent to the human body. For instance, typical studies on the thermal distribution within a human head, when exposed to signals emitted from mobile devices were found to raise the temperature by around 0.1 °C within the head.

2.1.2 Non-thermal Effects

According to the Institution of Electrical and Electronic Engineering (IEEE) standards C95.1, C95.3, and later updates [8], the non-thermal effects are defined as "any effect of EM energy absorption not associated with or dependent on the production of heat or a measurable rise in temperature."

The initial step towards assessing such effects is to determine whether exposure to EM waves can cause ionization within biological materials. Ionizing radiations are those providing enough energy capable of removing an electron (ionizing) of an atom or a molecule causing the chemical bonds within the genetic molecules (DNA) to break.

Fundamentally, the absorption of electromagnetic radiation is quantified in terms of energy quanta (hf), where f is the frequency and h is Plank's constant $(\sim 6.626 \times 10^{-34}\,\mathrm{J\,s})$. The energy quanta of RF radiation is measured in electron volt $(\mathrm{eV} \approx 160.217 \times 10^{-21}\,\mathrm{J})$, which is the *work* required to move an electron through a potential difference of one volt.

Based on the definition earlier, the energy quanta resulting from exposure to RF radiations in the frequency range 300 Hz–300 GHz are estimated to vary from 1.24×10^{-12} to 1.24×10^{-3} eV, respectively. Both of these values are extremely small in comparison with the energy required to break the weakest DNA chemical bond (about 1 eV). Hence, the energy associated with mobile phone radiations is considered to be non-ionizing.

Non-thermal biological effects can still exist within these energy levels. These effects can be detected if the effect of the electric field within the biological system exposed to RF fields is not masked by thermal noise (or random motion, or known as Brownian motion). Brownian motion exists due to the thermal energy that all objects possess at temperatures above absolute zero.

The averaged power of the thermal energy is determined by the relationship (kT_a), where k is Boltzmann's constant (86 µeV per degree), T_a is the absolute temperature measured in Kelvins ($T_a = 273 + t$), and t is the temperature in degrees centigrade. The averaged body temperature is around 37 °C (310 K), leading to kT_a of around 26.7 meV. If this value is much larger than the energy of the motion produced by the electric field, any effect of the electric field would be masked completely.

This gives an indication of the thermal noise existing within the living body and provides a good measure of the minimum electric field required to produce a detectable biological effect. In some cases, biological systems are more sensitive to some frequencies than others. This is because the thermal motion taking place at frequencies close to the resonance frequency of that biological system.

Ions existing within biological tissues are driven back and forth by the oscillating electric fields introduced externally from the source. The motion of these ions is severely reduced by the viscosity of the surrounding liquids. It has been argued that the movement of ions introduced by an electric field of 100 V/m is in fact less than 10^{-14} m (the diameter of an atomic nucleus). It was also argued that non-thermal biological effects are not necessarily caused by this ionic motion for cells whose radii are smaller than 10 µm. However, larger cells with greater mass were found to have higher influence due to the increased sensitivity to externally applied electric fields.

Attraction between ions could, however, be influenced by the existence of an electric field. The E-field polarizes the cell and makes it act as an electric dipole. This results in a polarized cell attracting similar polarized cells.

Biological effects associated with cell membranes can also exist at RF exposure conditions. These effects appear due to the electrical current flow introduced through the membrane in either direction. Membranes are known to have strongly non-linear electrical properties; hence, the voltage applied across the membrane is not always proportional to the current induced. Part of this non-linearity is due to the effect of the E-field on the proteins in the membrane area, which assists the product currents flowing through the membrane. In addition, membranes are involved in active chemical reactions that selectively change their porosity to various ions so that both electrical potentials and chemical signals may change the membrane's conductivity by orders of magnitude.

The interaction of EM fields with the body occurs on several levels, such as coupling penetration of fields, induction of fields, resonance, and others. These mechanisms and others result in a variation of the strength and distributions of the internal fields. Examples of these parameters are known as the body's electrical properties, shape, size, and orientation, as well as the strength and type of exposure.

In reality, electromagnetic exposure occurs in non-idealistic environment and humans are subject to exposure from different orientations and various distances from RF sources. Furthermore, humans (or animals) themselves are not identical in size and their dielectric properties vary from one to another, especially during their various growth stages [9, 10]. These facts are also applicable to humans, and correspond well to the concerns raised about the possibility that children might be affected under RF exposure.

2.2 Physical Models to Represent the Interaction of EMFs with Biological Tissue

Due to the complex nature of the biological matter, it is more viable to assess the EM behavior within the exposed tissue numerically, or by relying on measurements within simplified physical models, in the form of solid or liquid phantoms. In all assessment cases, measurements are typically supported with the detailed analysis of the physical interaction mechanisms which occur at different levels the microscopic and macroscopic levels.

2.2.1 Interaction Mechanisms

Interactions of tissues with EM fields are based on several established mechanisms for each of the electric and magnetic field components. The electric fields are associated with forces on the presence of electric charges, whereas the magnetic fields exist as a result of the physical movement of electric charges (electrical currents).

Under EM field exposure, the time varying electric field results in the re-orientations of the dipoles within the tissue, altering tissue bound charge orientation. This process creates new dipoles and also creates a flow of electric charges that forms the electric current flow.

The time varying magnetic field components are, however, responsible for inducting internal electric fields and circulating electric currents. The magnitudes of dipole polarization, charge induction, and flow are related to the tissue properties (permittivity and conductivity), the strength and polarization of the applied field, and the dimensions of the body.

Materials exposed to EM field may experience variations within their intermolecular structure. The applied E-field causes displacement of the material's free and bound electrons and magnetic fields result in orientation of atomic moments. Local fields inside materials are different from the applied fields because of the effect of charges of the surrounding molecules. Forces exerted by the applied

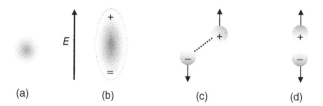

Figure 2.1 An illustration of the effects of the applied electric field (E) on the bound charges, dipole orientation, and drift of conduction charges. Each figure corresponds to (a) no electric field applied; and the effect of E-field on (b) separation of bound charges, (c) orientation alignment of permanent electric dipoles, and (d) drifting effects of free charges in response to the applied E-field.

electric and magnetic fields may change charged particles within the exposed material. This results in altering the material's charge patterns. This phenomenon could also cause additional electric and magnetic localized field.

2.2.1.1   Effects of Bound Charges

Bound charges are tightly held within their material and they are involved in limited movements around their original positions. Normally, positive and negative bound charges are superimposed upon each other cancelling their net charge (Figure 2.1a).

Applying an E-field that opposes the bound charges causes their separation and creates induced dipoles consisting of a combination of positive and negative charges separated by a small distance (Figure 2.1b). The new induced fields can be created from the polarization charge created by altering the balanced net charge.

2.2.1.2   Effects of Dipole Orientations

Most materials consist of molecules made of charge arrangements forming permanent dipoles. Orientations of these dipoles are random within the material due to the thermal excitations. When an E-field is applied, the resulting force on the permanent dipole tends to align the dipole with the applied E-field (Figure 2.1c). Again, the net alignment of the permanent dipoles results in new induced fields depending on the strength of the applied field.

2.2.1.3   Drift of Conduction Charges

When charges such as electrons or ions are free enough within the exposed material, the applied E-field can cause significant movement of these charges forming induced currents (Figure 2.1d). New fields are hence produced from the currents resulted from the sums of drifting conduction charges. In general, electric fields that are external to the body induce surface charges and the drifting of these induced charges forms currents on the body.

2.2.2 Dielectric Properties of Biological Materials

Generally, biological materials consist of a complex mix of water, ions, polar molecules, proteins, lipids, hormones, and many others. Characterization of the dielectric properties of such complex materials is heavily dependent on their actual composition and the frequency of the applied field. Effectively, the interaction of EM fields with such materials occurs on different levels depending on the size and the properties of the molecules.

2.2.2.1 Relaxation Theory

The dielectric properties of materials are generally defined in terms of their relative permittivity (ε') and conductivity (σ). These two parameters, respectively, represent the charge and current densities induced in response to an applied electric field of unit amplitude. Naturally, the dielectric properties of materials can be affected by several factors, such as frequency or temperature. Material dielectric properties are normally defined in the form of complex permittivity (ε^*) as shown in the following equation:

$$\varepsilon^* = \varepsilon_0(\varepsilon' - j\varepsilon'') \ (\text{F/m}) \tag{2.3}$$

The real part (ε') is known as the relative permittivity (or dielectric constant), which reflects the amount of polarization produced by the electric field. The imaginary part ε'' is known as the "loss factor" and is related to the conductivity of the material (σ) as shown in Eq. (2.4) [11]. The material conductivity (σ) is related to the displacement conductivity (σ_d), whereas in biological tissues the conductivity is given by the following equation:

$$\sigma = \sigma_d + \sigma_i \ (\text{S/m}) \tag{2.4}$$

where σ_i is the ionic conductivity due to the drift of free ions under the action of the field in biological tissue. The conductivity (σ) is related to the loss factor in terms of the absolute permittivity (ε_0) and the angular frequency ($\omega = 2\pi f$), as shown in Eq. (2.5):

$$\varepsilon'' = \frac{\sigma}{\varepsilon_0 \omega} \tag{2.5}$$

In the case of biological tissues, ε'' represents the ionic conductivity and the absorption due to processes of relaxation, including the friction associated with the alignment of the electric dipoles and with vibration and rotation in molecules. In principle, the response of any material to a voltage variation involves physical displacement of charge and the kinetics of this displacement establishes the frequency response of the bulk properties.

In nature, such variations could result in either relaxation or resonance forms. Generally, relaxation responses for most materials can be described by the physical processes involved, which can be represented in terms of the relaxation time constant (τ). However, there are some materials whose dielectric processes are known to have more than one relaxation time constant.

Furthermore, the complex physical nature of biological materials allows several relaxation processes to take place simultaneously. Hence, the total electrical response of the material can be characterized by several time constants.

The Debye relaxation theory has been typically used to characterize the dielectric properties of materials in terms of the relaxation time (τ). First and second order Debye equations are illustrated in Eqs. (2.6) and (2.7), respectively [12].

$$\varepsilon^{*}_{(\omega)} = \varepsilon_{\infty} + \frac{\varepsilon_{s} - \varepsilon_{\infty}}{1 + j\omega\tau} \tag{2.6}$$

$$\varepsilon^{*}_{(\omega)} = \varepsilon_{\infty} + \frac{\varepsilon_{s1} - \varepsilon_{\infty}}{1 + j\omega\tau_{1}} + \frac{\varepsilon_{s2} - \varepsilon_{\infty}}{1 + j\omega\tau_{2}} \tag{2.7}$$

where ε^{*} is the complex permittivity, ε_{s} is the relative permittivity at low frequencies (static), and ε_{∞} correspond to the permittivity at high frequencies (optical).

Characterization of the dielectric relaxation spectrum of biological materials extends over the EM spectrum with different dispersion regions which are dependent on the type of biological material and the exposure frequency. Biological tissues in general are characterized by three main dispersion regions (α, β, and γ), which represent the relaxation processes for the frequencies ranging from a few Hertz to a few Giga-Hertz [13] (see Table 2.1).

For frequencies up to a few tens of kilohertz, ionic diffusion processes at the site of the cellular membrane introduces dispersions in the α region. Polarization of the cellular membrane and organic macromolecules for the frequency region (10 kHz–100 MHz) bring the β dispersion region into effect.

In the gigahertz region, the γ dispersion exists due to the polarization of tissue water molecule. Between β and γ regions there exists a minor relaxation region known as δ. The δ region is caused partially by the rotation of amino acids, by the partial rotation of charged groups of proteins and the relaxation of protein-bounds.

Table 2.1 Frequency characterization of the main biological tissue relaxation regions.

Region	Frequency range (Hz)
α	1–10^{4}
β	10^{4}–10^{8}
δ	10^{8}–10^{9}
γ	$\sim 2 \times 10^{10}$

In general, the permittivity (ε) values decrease as the frequency increases. This is due to the inability of the charges in tissue to respond to higher frequencies where the relaxation response becomes limited to the polarized and permanent molecular dipoles (e.g. water).

2.2.2.2 Age-Dependent Dielectric Properties

Aging is generally characterized by the increase in size of the body and organs, by variations of their tissues dielectric properties as a direct consequence of their changes in the cellular structure and composition of the body and by their water concentration as well. For this reason, the dielectric properties of the body organs and tissues tend to vary at each growth stage. Such variations cannot be assumed to change linearly with age and are not applicable to all tissue types within the body [10]. This study has also shown that there is a general trend of decreasing permittivity and conductivity with increasing age for most tissue types. Furthermore, these changes in dielectric properties were found to be more pronounced in certain tissue types, such as the brain, the skin, or the skull tissues. Projection of these results to human tissues suggests disparity in the absorption of RF radiations by younger people in comparison to adults.

During the past two decades, a comprehensible characterization of tissue dielectric properties of humans and animals have been conducted, covering different age developmental stages. The frequency- and age-dependent tissue dielectric properties raised questions about the significance of these findings on the EM field interaction with adults and children.

2.2.3 The Interaction of EM Fields with Biological Materials

The dielectric properties of biological materials can be considered as one of the many factors that determine the interaction with EM fields. These properties are generally used to characterize the bulk EM property of a given material sample and the effect that might initiate biological responses.

In the case of body exposure to RF, the incident fields exposed on the body undergo various measures of propagation and may suffer various boundary effects such as reflection or scattering depending on tissues' properties, shape, size, direction, and the angle of exposure.

Coupling between EM fields and the exposed body may exist in various amounts depending on the field characteristics, the body dimensions and material properties. The maximum coupling can be achieved when the size of body is of the same order of magnitude as the wavelength and when the long axis of the exposed body is in the direction and polarization of the electric field. Internal electric fields inside the body are a combination of the penetrated fields and those induced from the external fields. These fields may also interact with bound- and free- charges, forming ionic currents which may interact with the local fields at the cell level, in the extra-cellular space, within cells, and across cell membranes.

Tissues exposed to electric and magnetic fields by a nearby antenna might be subject to both thermal and non-thermal biological implications (Section 2.2). There are several mechanisms by which energy can be transferred from the EM fields into a biological material. For instance, energy can be gained by the electric charges within the biological material due to the forces exerted by the applied electric fields.

The magnetic fields, however, do not pass energy to charges, but the forces that the magnetic fields exert on the charges can change their direction. Nevertheless, the magnetic fields can convey energy through the forces they exert on the magnetic dipoles within the material. Clearly, the transfer of energy from the magnetic fields is less pronounced in biological tissues due to the fact that biological materials are predominantly non-magnetic.

Although biological materials are generally treated as non-magnetic (their permeability is essentially equal to that of free space), magnetic fields are still thought to affect living species in many ways. For instance, it is well established that static magnetic fields are used by animals, such as birds, for tracing directions. From this, the effects of magnetic fields are likely to be most sensitive in biological tissues containing small amounts of magnetite (for static cases). In the case of time-varying magnetic fields, the most evident way is the induction of electric fields from the magnetically induced currents (within tissues).

However, small amounts of magnetite molecules exist within human tissues as naturally occurring oxide of ferrimagnetic iron (Fe_3O_4). This type of ferrimagnet behaves similarly within magnetic fields to a ferromagnet such as iron. Under exposure to EM fields which are limited by the current guidelines, it has been calculated that the largest RF effect that could be possibly caused by magnetic fields is extremely small. Any other effects due to magnetic fields at these frequencies should be even smaller.

The previous text suggests that biological effects from EM fields are much more likely to be caused by electric rather than magnetic fields. EM field interaction with biological tissues is influenced by many factors such as the radiation frequency and shape, size, and the dielectric properties of the material under exposure.

2.2.3.1 Interactions on the Body Scale

Interactions of the EM fields in the communication frequency band with the body can be assessed in terms of the whole-body averaged and localized EM energy patterns. The wavelength of this particular frequency band can be considered to be comparable to the size of the exposed body. Coupling and resonance tend to dominate the extent of interaction mechanisms between the fields and the body.

In nature, the organs within the body have various shapes and sizes and consist of different tissue types. For this reason, the peak EM distributions within the exposed body tend to vary significantly with frequency. Organs such as the brain, eyes, or ears may particularly undergo higher EM absorptions as a result of exposure. Peak power absorptions and EM resonance within the organs are particularly important in health and safety risk assessments.

2.2.3.2 Interactions on the Tissue Scale

At the tissue scale level (e.g. in vitro samples), the size of the exposed sample is considerably smaller than the wavelength of the communication frequency band. In this case, propagation mechanisms tend to dominate and the biological organ is simply seen by the EM fields as a scatterer. Hence, biological tissue sample can be represented in the form of a homogeneous lossy dielectric medium.

2.2.3.3 Interaction on the Cellular and Sub-cellular Scales

At the microscopic level, the wavelengths of the communication frequency band are considered to be much larger than the exposed objects. Analysis in such conditions can be carried out using a "quasi-static" approximation approach, where the spatial variation of the E and H fields are approximated as being the same within that of static EM fields.

The electric-circuit theory can also be applied as a simpler approach in comparison with the EM solution. In this method the voltages and currents are the principle variables. Hodgkin and Huxley (HH) [14] have established an equivalent circuit for describing the membrane complex electrical activity in terms of circuit theory. Several other improved models have been developed over the years describing the operational mechanisms of an ionic channel, such as a finite-states Markov's machine [15]. Further microscopic studies have been attempted in order to investigate the effect of EM field interactions with an ion and a receptor site based on the quantum Zeeman–Stark model [15, 16].

2.3 Dosimetry Concepts

In order to understand the effects of EM fields on biological tissues, it is necessary to determine the magnitude of the exposed fields within the various parts of the object (or body) [17]. This requires knowledge of the dielectric properties of the different types of tissues and the signal type and field strengths and many other parameters. Assessments of the EM energy absorption within a given mass of tissue can be carried out using the averaged field strength together with the tissue electrical properties at that given location. Several dosimetric parameters have been used to assess the exposure to biological systems, such as the maximum permeable exposure (MPE) and the more commonly used, specific absorption (SA) and SAR.

2.3.1 The Specific Absorption Rate (SAR)

The SAR is used in dosimetry to denote the transfer of energy from the EM fields to biological materials (rate of energy deposition per unit mass of tissue). The SAR at a point is defined as the rate of change of energy absorbed by charged particles within an infinitesimal volume at that point within an absorber, averaged by the mass of that small volume.

$$\text{SAR} = \frac{\partial W/\partial t}{\rho_m} \ (\text{W/kg}) \tag{2.8}$$

where ρ_m is the mass density.

However, the rate of change of energy $\partial W/\partial t$ is equivalent to power density (P). Hence, Eq. (2.8) can be rewritten as

$$\text{SAR} = \frac{P}{\rho_m} \ (\text{W/kg}) \tag{2.9}$$

The previous can be re-formulated to relate the SAR to the internal E-fields:

$$\text{SAR} = \frac{\sigma |E|^2}{\rho_m} \ (\text{W/kg}) \tag{2.10}$$

where σ represents the conductivity of the material, $|E|$ is the rms magnitude of the electric field and ρ is the mass density.

The whole-body averaged SAR is defined as the time rate of change of the total energy transferred to the absorber body, divided by the total mass of the body.

In the IEEE standard C95.3, the SAR is defined as the time derivative of the incremental energy (dW) dissipated in an incremental mass (dm) contained in a volume element (dV) of a given density (ρ).

$$\text{SAR} = \frac{d}{dt}\left(\frac{dW}{dm}\right) = \frac{d}{dt}\left(\frac{dW}{\rho \cdot dV}\right) \ (\text{W/kg}) \tag{2.11}$$

The SAR is also related to heat capacity in terms of the rate of change of temperature within the exposed object as shown as follows:

$$\text{SAR} = c \cdot \frac{\Delta T}{\Delta t} \ (\text{W/kg}) \tag{2.12}$$

where c is the specific heat capacity, ΔT is the change in temperature, and Δt is change in time. This equation is based on the assumption that any heat transportation mechanisms are negligible.

2.3.1.1 SAR Measurement Techniques over the Frequency Spectrum

Different measurement techniques are needed to evaluate the SAR induced at the exposed body over the frequency spectrum. At frequencies below 100 MHz, the induced current flowing through the body towards the ground is measured in order to assess the induced SAR. Below a few hundreds of megahertz, SAR values can be determined by measuring both of the electric and magnetic fields.

Assessment of the SAR in the frequency band (300 MHz–3 GHz) can be achieved by measuring either the temperature or electric field distributions within the exposed object (Section 2.4.1).

For exposure to frequencies below few Gigahertz, the SAR values can also be evaluated by measuring the thermal profiles using RF-transparent temperature sensors. Above few Gigahertz, most of the EM absorption occurs on the surface of the lossy material, and hence thermographic cameras can be used to measure the SAR at the surface of the exposed body.

2.3.1.2 SAR Spatial Averaging

Measurements and computations of the SAR values are typically carried out through averaging data over volumes containing a given tissue mass (e.g. 1 g). The human body consists of a vast number of tissue types forming regions (or organs) of different shapes, sizes, and mass densities. Averaging data over a fixed volume is therefore not possible over an inhomogeneous volume of tissues. Other difficulties could rise when discretizing the body for computation, e.g. FDTD gridding, where the voxel does not necessary hold the averaging mass required. This problem is addressed in the IEC/IEEE 62704-1 standard, which recommended procedures for SAR spatial averaging.

The IEEE averaging procedure recommends considering a tissue sub-volume such that it does not extend exterior surfaces of the body, but may have pockets of air within it. The mass of each sub-volume may not be smaller than the averaging mass (e.g. 1 g) but preferably as close to it as possible. Averaging of the SAR over a fixed mass of tissue in the body should be considered separately for each type. Any other tissue type contained within the cube volume should be treated as air (mass = 0 and SAR = 0).

2.3.1.3 Tissue Mass Averaging Procedures

According to the IEEE recommendation, the peak spatial-averaged SAR should be assessed over cubical tissue volumes containing a mass within 5% of that required (e.g. of the 1 or 10 g).

The peak spatial-averaged SAR values are commonly averaged over 1 g of tissue volume, whereas 10 g averaging is used for extremities cases (e.g. peak SAR values at hands).

The size of the cubical volume centered at each location is increased from all cube sides until the desired value for the required mass should be accomplished without extending beyond the most exterior surface of the body. If the required tissue mass was not obtained, the center of the cubical volume should be moved to the next location. The centers of each valid averaging volume should be assigned with the averaged SAR values. All locations included as part of this averaging volume should be flagged so as to indicate that they have been averaged at least once, whereas locations that are not part of any valid averaging volume should be marked as "unused."

Any positions marked as "used," and have not been in the center of any averaging volume should be assigned the highest mass averaged SAR value of that averaging volume. In the case of an "unused" location, a new averaging volume is constructed with the unused location centered at one surface of the averaging cube and the other five surfaces expanded evenly in all directions until the required mass of tissue is obtained.

For cases where averaging is performed over tissue volumes that contain mass less than that used for averaging (e.g. 10 g averaged peak SAR of ear pinna), the average SAR should be determined as the total SAR averaged over the total mass of the tissue of concern (Figures 2.2–2.4).

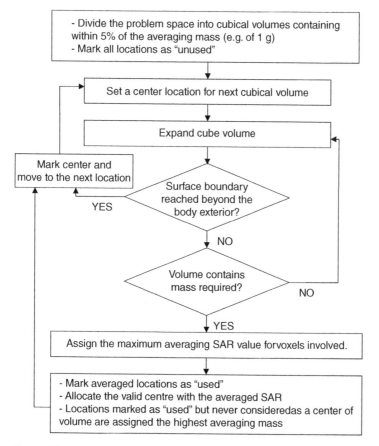

Figure 2.2 Flowchart showing the basic IEEE procedure for peak SAR numerical averaging.

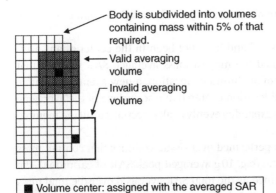

Body is subdivided into volumes containing mass within 5% of that required.

Valid averaging volume

Invalid averaging volume

Figure 2.3 A valid setup for the SAR averaging volumes. Based on IEC/IEEE International Standard – IEC/IEEE 62704-1: 2017, pp. 1–86, 27 Oct. 2017, doi: 10.1109/IEEESTD.2017 .8088404.

■ Volume center: assigned with the averaged SAR
▨ Voxels marked as "used" within the averaging

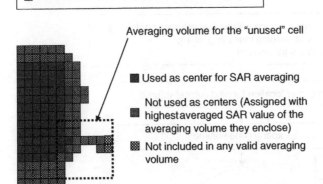

Averaging volume for the "unused" cell

■ Used as center for SAR averaging

▨ Not used as centers (Assigned with highest averaged SAR value of the averaging volume they enclose)

▩ Not included in any valid averaging volume

Figure 2.4 Assignment of SAR values for each tissue sub-volume. Based on IEC/IEEE International Standard – IEC/IEEE 62704-1: 2017, pp. 1–86, 27 Oct. 2017, doi: 10.1109/ IEEESTD.2017.8088404.

2.3.1.4 Localized and Whole-Body Averaged SAR

There are special cases where the averaging mass is very small and hence the average SAR can be expressed as the total SAR averaged over the total mass of the body part of concern. The whole-body averaged SAR is therefore defined as the total absorbed power averaged over the total absorbing body mass.

2.3.2 The Specific Absorption (SA)

Characterization of the interaction of EM fields with biological matter is commonly evaluated using a dosimetric parameter known as the SAR. However, the situation is different for exposure to pulsed fields, and another dosimetric parameter is needed. Several studies have utilized the SA, which is defined as the quotient of the incremental energy (dW) absorbed by an incremental mass contained in a volume of a given density $(\rho \cdot dV)$.

Evaluation of the SA of biological materials in the time domain is a complicated issue. Although many techniques are available for solving dispersive materials in the time domain, computation of the SA involves further complexity, especially when integrating the multiple of the dispersive tissue conductivity by the local E-fields over the pulse duration. For this reason, most of the previous reported results are based on assumptions of constant conductivity value [18, 19].

2.4 Dosimetry Methodology

Dosimetry is generally defined as the fundamental process of quantifying the EM energy absorbed inside biological materials. Typically, dosimetric data are used to determine whether biological responses could be induced due to non-ionizing radiation exposure.

Theoretical, experimental and numerical methods can be used for dosimetry. Each of these dosimetry approaches is preferred for different scenarios (Table 2.2).

This section highlights some numerical and experimental dosimetry methods. By inspecting this table it can be seen that with the development of advanced computation methods, combination of experimental dosimetry with numerical modelling is becoming more desirable.

2.4.1 Experimental Dosimetry

Measurement provides the means to describe physical phenomena in quantitative terms. So far, most of the SAR compliance tests are carried out by using a sensitive electric field probe to scan a certain volume inside a phantom, filled with a liquid simulating tissue-equivalent material. Such an experimental setup enables

Table 2.2 A comparison of the current numerical and experimental electromagnetic dosimetry techniques.

Technique	Numerical dosimetry		Experimental dosimetry	
	Theoretical	Modelling	Probe measurement	Thermal imaging
Surface SAR	Simple models only	Yes	No	Yes
Volume SAR	Simple models only	Yes	Yes	No
Multi-tissue	No (Complicated)	Yes	No	Yes
Typical application	Averaged shapes	Detailed models	Liquid phantoms	Solid phantoms
Method	—	—	Invasive	Non-invasive

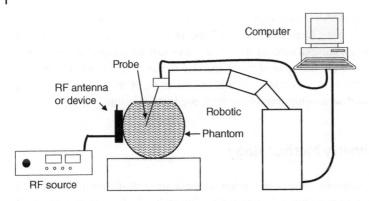

Figure 2.5 Schematic diagram of the probe field measurement within liquid phantom.

dosimetry studies to detect an approximate peak SAR distribution patterns and averaged SAR values (Figure 2.5).

This method has been used as a standard technique for SAR evaluations despite the fact that it takes a long measurement time due to the scanning of large number of points. The utilization of the probe makes this method invasive and limits its function to the liquid phantom. Furthermore, such a method cannot be used to measure the SAR values on the surface of the phantom, for example, the ear region of the phantom head where the peak SAR is usually located.

Thermal imaging on solid phantoms is a non-invasive method which has potential of overcoming the limitations of the probe technique. A thermographic camera is used to detect the thermal image on a cross section of the phantom under RF radiation. The local SAR is then determined by making a linear extrapolation from the rate of temperature change during the linear portion of the heating curve (Eq. 2.12).

The main advantage of utilizing solid phantoms is that it enables non-invasive measurements on physical multi-tissue models. However, it involves further complexity and does not provide direct measurement of the SAR within a given volume. Recent advances of the solid phantom techniques have encouraged a number of researchers to select this method for analysis (Figure 2.6).

2.4.2 Numerical Dosimetry

Numerical dosimetry involves computing the EM wave propagation through discretized anatomic models of biological bodies and assessment of their SAR distributions.

2.4.2.1 Theoretical Analysis
Prior to the development of fast computational electromagnetic methods, theoretical dosimetry has been used extensively to approximate the internal fields and

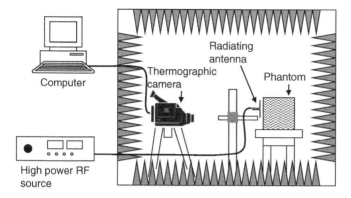

Figure 2.6 Schematic diagram of the thermographic measurement set-up.

EM energy absorption within simple basic volumes. Several studies have established certain simplified techniques for evaluating the EM interaction with biological objects by means of approximating them with cubical, spheroid, or spherical volumes. This type of theoretical assessment is still being used extensively for evaluating the interaction of EM fields with biological materials at the cellular and sub-cellular scales [20, 21].

Many theoretical and empirical techniques have also been applied to evaluate the interaction of certain EM-radiation scenarios with simplified objects. Theoretical dosimetry has also been used extensively in conjunction with experimental or numerical studies, particularly for cellular level investigations and for relating the surface SAR data to volumetric SAR distributions (or vice versa).

2.4.2.2 Numerical Modelling

Evaluation of the internal fields within any EM-radiated object can be carried out by solving Maxwell's equations using various numerical techniques.

The SAR is assessed using the local EMFs within the lossy-volume of biological tissues (Section 2.4.1). In some occasions, SAR distributions are used to evaluate the energy absorptions and the thermal profiles within the structure. Thermal assessments can be achieved by using the Bioheat transfer functions.

EM-based numerical modelling has become increasingly viable due to the recent advances in both hardware and software aspects of computational resources. The finite difference time domain (FDTD) method has particularly grown in importance over the last few years as a time-domain solver for EMFs of arbitrary three-dimensional structures [18–22].

The importance of FDTD has evolved as a result of its ability to discretize Maxwell's curl equations for solving complex EM problems with reasonable computational requirements. In this context, it offers the feasibility of achieving accurate characterization of complex inhomogeneous structures that could possibly consist of inhomogeneous materials with various dielectric or magnetic

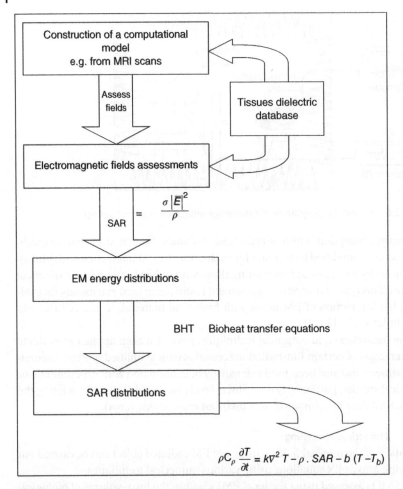

Figure 2.7 Computation procedure for computing the EM, SAR, and thermal distributions, which are induced as a result of EM field interactions.

properties. Currently, several models are made available for high-resolution computer-suited anatomical structures, representing the various organs and tissue types (Figure 2.7).

2.5 Numerical Dosimetry at the Radiofrequency and Microwave Regions

Numerical dosimetry computations are typically carried out to evaluate the whole-body and peak SAR distributions inside the body when exposed to devices emitting EM waves. The numerical approach allows researchers assess the interactions of EMF with the human body, accurately and efficiently. The numerical approach also enables evaluating the effects of varying the electric, magnetic, and propagation vectors are aligned along the long axis of the body alternatively.

Determination of the SAR can be attained numerically or experimentally depending on the complexity of the problem, and on the type of the study. In the case of inhomogeneous anatomical body models, it is rather complicated and difficult to perform the physical SAR measurements. For this reason, numerical dosimetry is the most feasible solution for this problem. The SAR profiles within the body can be determined by assessing the EM field distribution within the model using computational methods, such as the FDTD method or other equivalent methods such as the finite integral technique (FIT), in conjunction with the tissue properties.

The next few sections provide detailed summary of time-domain computation methods which are most commonly used in SAR evaluations.

2.5.1 Formulation of the Scattered-Field FDTD Algorithm

The FDTD technique has been widely used in this research area due to its strengths in solving Maxwell's equations accurately and efficiently within a highly scattering medium. It can be used to calculate either the scattered or total fields within a given computational domain.

The total-field computation approach is also known as the classical Yee approach for solving the EM fields. The scattered field method, however, is based on solving the Maxwell's equations for the scattered fields separately from the incident fields. In this method, the incident fields are computed using the classical FDTD approach whereas the local incident-field is defined as a function of a known source and the medium material parameters. Computations of the resulting fields within the region of interest are simply achieved by adding the incident and scattered fields for each cell (Eq. (2.13)) [22–24].

$$E_{\text{Total}} = E_{\text{Incident}} + E_{\text{Scattered}}, \text{ and } H_{\text{Total}} = H_{\text{Incident}} + E_{\text{Scattered}} \tag{2.13}$$

The elimination of the incident fields from the computational domain enables lower reflection levels by reducing the computational load on the outer absorbing boundary conditions (ABCs).

The most common implementation is the split-field FDTD equations in order to improve the computation accuracy by applying the perfectly matched layer (PML) ABC. A full description of the program operation is shown in Figure 2.8.

In order to validate the computational approach of this study, the main FDTD program structure is tested and compared to a commercial code based on solving Maxwell's equations using the FIT.

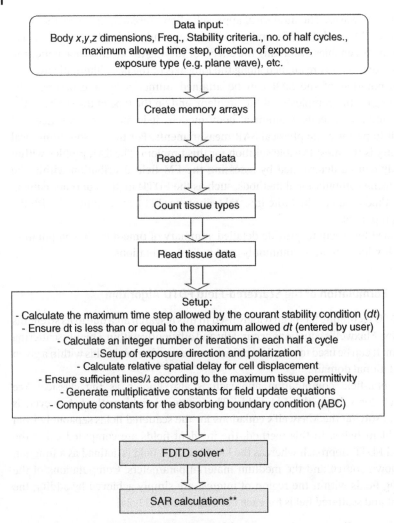

Figure 2.8 Operation of the FDTD program.
* The FDTD solver iteration chart is shown in Figure 2.9.
** Computed for each voxel according to Figure 2.9. Organ averaged SAR values are achieved by dividing the total absorbed power by the total mass for each tissue type.

2.5.2 Discretization of Anatomical Models in FDTD

Numerical computations of the interaction of EM fields with the body aim to mimic the anatomy of the body. Typical numerical models representing different tissue types can be generated based on digitizing magnetic resonance images (MRI) into volume cells (voxels). Discretization of the model was simply achieved

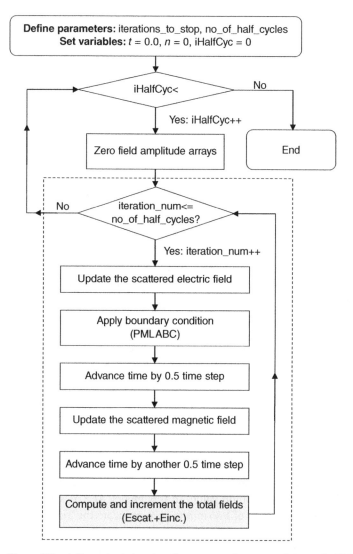

Figure 2.9 A Flow chart showing the computation steps of the main FDTD program.

by creating color-coded images of the different organs and tissue types in the form of slices along the body.

The color-coded images slicing through the anatomical rat were combined into a file containing data representing a three-dimensional FDTD human model of the different tissue types.

The dielectric properties of each tissue type within the body are accessed from a separate file which contains a database of the measured age-dependent and dispersive dielectric data (Figure 2.10).

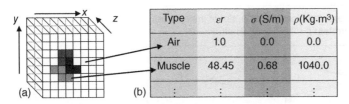

Type	εr	σ (S/m)	ρ(Kg·m³)
Air	1.0	0.0	0.0
Muscle	48.45	0.68	1040.0
:	:	:	:

Figure 2.10 The file structure for (a) model voxel configurations; and (b) tissues properties.

2.5.3 Comparisons of Numerical Results with Analytical Benchmarks

A theoretical calculation model for expressing the EM field and SAR induced inside a lossy dielectric sphere by an incident plane wave has been used as a benchmark for validating the developed FDTD software.

The analytical solution was based on exposing single-layered muscle-equivalent phantom spheres, with radii of 3 and 15 cm, to a plane wave of 1000 MHz radiation [25]. Similar dielectric properties of muscle tissue were used in both analytical and computational models [26].

In this section, the theoretically predicted and computed SAR patterns are compared based on the patterns along the X, Y, and Z directions.

Figure 2.11 show the analytically solved SAR patterns along the three major rectangular axes X, Y and Z of the 3 cm sphere [25]. For the convenience of presentation, the SAR values are normalized to the maximum SAR given for each figure. It was found that strong standing wave patterns (hot spots) exist in the sphere. This absorption characteristic of RF radiation may be attributed to the focusing effect of the EM field inside a dielectric body when the size of the body is comparable to the wavelength of the radiation inside the body.

Figure 2.12 illustrates the computed SAR patterns computed inside the 3 cm sphere using the developed FDTD program (with PML ABC). The SAR values have also been normalized to the maximum value. In this model, $41 \times 41 \times 41$ voxels have been allocated with cell sizes of 1.46 mm to accommodate the 3 cm sphere. This configuration satisfies a dense meshing of 29 cells per wavelength (inside the dielectric). The sphere models have been separated from the PML layers by 10 cells representing free space in order to ensure better ABC performance. One of the disadvantages of fixing the number of surrounding free-space cells is the variability in separation distance between the tested object and the ABC, specifically when varying the size of the cell.

Based on these figures, an excellent correlation has been observed among the overall SAR pattern which have been plotted from the analytical predictions and FDTD computations.

Figure 2.13 shows the analytically solved SAR patterns along the three major rectangular axes X, Y, and Z of the 15 cm sphere [25]. Again, for the convenience of

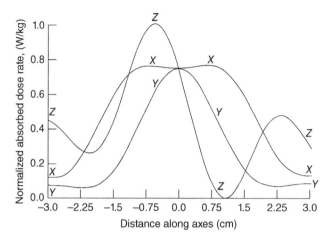

Figure 2.11 Plots of the normalized analytically assessed SAR patterns along the major axes of a muscle-equivalent dielectric sphere (radius = 3 cm). Calculations are based on a plane wave source of 1000 MHz, with energy flux density of 1 mW/m². Whole-body averaged SAR = 0.235 W/kg. Source: Reproduced from [25].

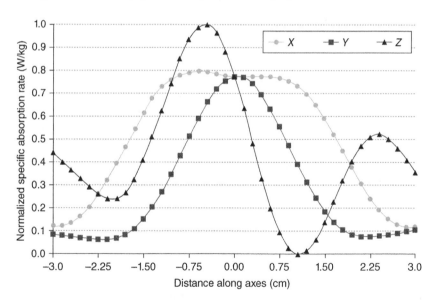

Figure 2.12 Plots of the normalized FDTD-computed SAR patterns along the major axes of a muscle-equivalent dielectric sphere (radius = 3 cm). Calculations are based on a plane wave source of 1000 MHz, with energy flux density of 1 mW/m² (computed whole-body averaged SAR = 0.218 W/kg). The sphere has been modelled using 41 × 41 × 41 voxels, surrounded by 10 voxels to separate the object from the PML boundary.

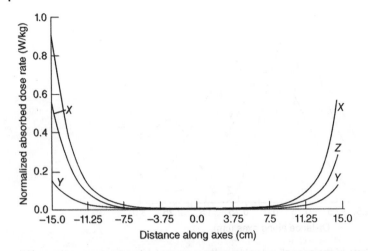

Figure 2.13 Plots of the normalized analytically assessed SAR patterns along the major axes of a muscle-equivalent dielectric sphere (radius = 15 cm). Calculations are based on a plane wave source of 1000 MHz, with energy flux density of 1 mW/m². Whole-body averaged SAR = 0.0365 W/kg. Source: Reproduced from [25].

presentation, the SAR values are normalized to the maximum SAR given for each figure. Unlike the 3 cm sphere, it was noticed that stronger absorptions occurred close to the surface of the 15 cm sphere.

Figure 2.14 illustrates the computed SAR patterns computed inside the 15 cm sphere using the developed FDTD program (with PML ABC). The SAR values have also been normalized to the maximum value. In this model, 41 × 41 × 41 voxels have been allocated with cell sizes of 7.32 mm to accommodate the 15 cm sphere. This configuration only satisfies 6 cells per wavelength (inside the dielectric), which could produce some numerical error.

The sphere models have been separated from the PML layers by 10 cells representing free space in order to ensure better ABC performance. One of the disadvantages of fixing the number of surrounding free-space cells is the variability in separation distance between the tested object and the ABC, specifically when varying the size of the cell.

Despite the dissatisfactory Courant criteria (only 6 cells per wavelength rather than 10), comparisons among the analytical and numerical computations of the SAR inside the 15 cm sphere have shown acceptable pattern correlation. However, the absorption patterns obtained using the FDTD program have suffered from several ripples. The maximum error was found to be around 20%.

Figure 2.15 illustrates the computed normalized SAR patterns computed inside the 15 cm sphere using the developed FDTD program (with PML ABC). In this

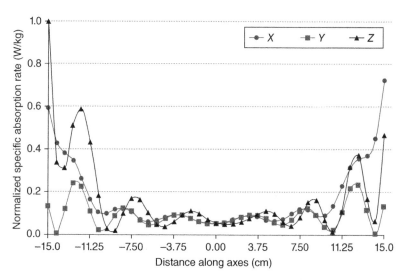

Figure 2.14 Plots of the normalized FDTD-computed SAR patterns along the major axes of a muscle-equivalent dielectric sphere (radius = 15 cm). Calculations are based on a plane wave source of 1000 MHz, with energy flux density of 1 mW/m² (computed whole-body averaged SAR = 0.0454 W/kg). The sphere has been modelled using 41 × 41 × 41 voxels, surrounded by 10 voxels to separate the object from the PML boundary. A significant error (about 20%) was caused by the insufficient Courant criteria (The cell size of this scenario only ensures 6 cells/λ).

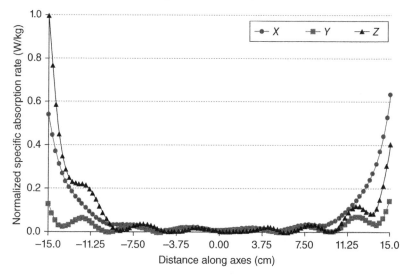

Figure 2.15 Plots of the normalized FDTD-computed SAR patterns along the major axes of a muscle-equivalent dielectric sphere (radius = 15 cm). Calculations are based on a plane wave source of 1000 MHz, with energy flux density of 1 mW/m². Whole-body averaged SAR = 0.0376 W/kg. The sphere has been modelled using 101 × 101 × 101 voxels, surrounded by 10 voxels for separation from the PML boundary.

model, $101 \times 101 \times 101$ voxels have been allocated with cell sizes of 2.97 mm to accommodate the 15 cm sphere. This configuration satisfies a dense meshing of 14 cells per wavelength (inside the dielectric).

Here, the ripples have been reduced significantly and an excellent correlation has been observed among the overall SAR plotted (Error <5%).

References

1 World Health Organization factsheets on EM fields. https://www.who.int/health-topics/electromagnetic-fields/.

2 ICNIRP, International Commission on Non-Ionizing Radiation Protection (2020). Guidelines for limiting exposure to electromagnetic fields (100 kHz to 300 GHz). *Health Physics* 118 (5): 483–524.

3 Majchrzak, E., Dziatkiewicz, G., Paruch, M. (2008). The modelling of heating a tissue subjected to external electromagnetic field. *Acta of Bioengineering and Biomechanics* 10 (2): 29–37.

4 Dimbylow, P.J. (1997). FDTD calculations of the whole-body averaged SAR in an anatomically realistic voxel model of the human body from 1 MHz to 1 GHz. *Physics in Medicine and Biology* 42: 479–490.

5 Dimbylow, P. and Bolch, W. (2007). Whole-body-averaged SAR from 50 MHz to 4 GHz in the University of Florida child voxel phantoms. *Physics in Medicine and Biology* 52: 6639–6649.

6 Pennes, H.H. (1948). Analysis of tissue and arterial blood temperatures in the resting human forearm. *Journal of Applied Physiology* 1 (2): 93–121.

7 Wang, J. and Fujiwara, O. (1999). FDTD computation of temperature rise in the human head for portable telephones. *IEEE Transactions on Microwave Theory and Techniques* 47 (8): 1528–1534.

8 IEEE (2013). IEEE Recommended Practice for Determining the Peak Spatial-Average Specific Absorption Rate (SAR) in the Human Head from Wireless Communications Devices: Measurement Techniques. IEEE Std. 1528–2013 (Revision of IEEE Std 1528–2003), pp. 1–246.

9 Peyman, A., Rezazadeh, A.A., and Gabriel, C. (2001). Changes in the dielectric properties of rat tissue as a function of age at microwave frequencies. *Physics in Medicine and Biology* 46: 1617–1629.

10 Peyman A. and Gabriel C. (2002). Variation of the Dielectric Properties of Biological Tissue as a Function of Age *Final technical report, Department of Health, UK.*

11 Peyman, A. and Gabriel, C. (2010). Cole–Cole parameters for the dielectric properties of porcine tissues as a function of age at microwave frequencies. *Physics in Medicine and Biology* 55: 413–419.

12 Foster, K.R. and Schwan, H.P. (1989). Dielectric properties of tissues and biological materials: a critical review. *Critical Reviews in Biomedical Engineering* 17 (1): 25–104.

13 Gabriel, C. (2003). Dielectric spectroscopy of biological material – a tool for basic science and biomedical applications, IOP.

14 Hodgkin, A.L. and Huxley, A.F. (1952). A quantitive description of membrane current and its application to conduction and excitation in nerve. *The Journal of Physiology* 117: 500–544.

15 Emili, G., Schiavoni, A., Francavilla, M. et al. (2003). Computation of electromagnetic field inside a tissue at mobile communications frequencies. *IEEE Transactions on Microwave Theory and Techniques* 51 (1): 178–186.

16 Bernardi, P., d'Inzeo, G., and Pisa, S. (1994). A generalized ionic model of the neuronal membrane electrical activity. *IEEE Transactions on Biomedical Engineering* 41: 123–133.

17 IFAC. *Dielectric Properties of Body Tissues*. http://niremf.ifac.cnr.it/tissprop/htmlclie/htmlclie.htm (accessed Feb. 2015).

18 Okoniewska, E., Stuchly, M.A., and Okoniewski, M. (2004). Interactions of electrostatic discharge with the human body. *IEEE Transactions on Microwave Theory and Techniques* 52 (8): 2030–2039.

19 Dawson, T.W., Stuchly, M.A., and Kavet, R. (2004). Electric fields in the human body due to electrostatic discharges. *IEEE Transactions on Biomedical Engineering* 51 (8): 1460–1468.

20 DeBruin, K.A. and Krassowska, W. (1999). Modelling electroporation in a single cell. I. Effects of field strength and rest potential. *Biophysical Journal* 77: 1213–1224.

21 San-Martic, S.M. and Sebastian, J.L. (2003). Electric field distribution in realistic cell shape models. *Journal of Electrostatics* 57: 143–156.

22 Holland, R. and Williams, J.W. (1983). Total-field versus scattered-field finite-difference codes: a comparative assessment. *IEEE Transactions on Nuclear Science* 30 (6): 4583–4588.

23 Kunz, K.S. and Luebbers, R.J. (1993). *The Finite Difference Time Domain Method for Electromagnetics*. CRC Press, ISBN: 0-8493-8657-8.

24 Berenger, J.-P. (1994). A perfectly matched layer for the absorption of electromagnetic waves. *Journal of Computational Physics* 114: 185–200.

25 Ho, H.S. and Guy, A.W. (1975). Development of dosimetry for RF and microwave radiation-II: calculations of absorbed dose distributions in two sizes of muscle-equivalent spheres. *Health Physics* 29 (2): 317–324.

26 Schwan, H.P. and Piersol, G.M. (1954). The absorption of electromagnetic energy in body tissues, a review and critical analysis, part 1. Biophysical aspects. *American Journal of Physical Medicine* 33 (6): 371–404.

3

Numerical Exposure Assessments of Communication Systems at Higher Frequencies

Muhammad Rafaqat Ali Qureshi, Yasir Alfadhl, and Xiaodong Chen

School of Electronic Engineering and Computer Science, Queen Mary University of London, London, UK

3.1 Introduction

With the enormous increase in internet users, demands for high data rate from wireless and portable devices dramatically increased from last few years. Wireless devices manufacturers are facing huge challenge how to overcome the global large bandwidth requirements. 4G mobile communication systems or formally called long-term evolution (LTE) enable systems operates under 6 GHz frequency band. There are well-established assessment guidelines for devices that operate under 6 GHz [1] and 10 GHz [2] to keep the exposure levels under the basic limits and to minimize the public concern about over exposures.

5G wireless mobile communication systems expected to use frequency bands at millimeter waves (mmWs) range [3]. The mmW frequency band ranges from 30 GHz to 100 GHz, which is a part of radio frequency (RF) spectrum. The possible use of these higher frequency bands imposes new challenges in terms of electromagnetic (EM) exposure assessment. The mmW radiations are non-ionizing and the major concern is heating of tissues at these frequencies. Due to lack of availability of assessment methods, governments currently rely on standards that developed before the exponential increase in use of wireless communication devices.

Several attempts made in the past to investigate the effects of mmW exposure on human body. Majority of previous studies investigates the thermal effect caused by exposure from mmWs frequencies [4–8]. These studies use different techniques such as thermocouple, magnetic resonance thermal imager and various analytical solutions to estimate the temperature variations within the human body. Theoretical analysis on the steady state temperature elevation using bioheat equation is also very challenging as it depends on many parameters such as thermal conductivity, blood flow, and thermal and tissues thickness parameters. A complete

Low Electromagnetic Field Exposure Wireless Devices: Fundamentals and Recent Advances, First Edition.
Edited by Masood Ur Rehman and Muhammad Ali Jamshed.

literature review has been done in references [9–11], which provide basic knowledge about the mmWs behavior in terms of transmission and reflection at the skin tissue interface and dielectric properties of skin are also provide over the wide frequency range.

Due to the shallow penetration depth in biological material at mmW frequencies, most of the energy is absorbed within few millimeters of the skin. This makes challenging for conventional E-field probe to estimate EM absorption on skin surface. Therefore, specific absorption rate (SAR) levels were found significantly higher in comparison with those at microwave frequencies with same incident power density (IPD) [9–11]. International guidelines providing agencies recommend to use power density (PD) as a basic restriction limit to reduce the excessive exposure at higher frequencies (above 6 [1] and 10 GHz [2]). But devices operating in near- or far-field do not take account for field distribution or power absorption in the tissues, they only consider density of power travelling towards the tissue [9]. Therefore, PD may not likely be useful as SAR or temperature for assessing safety at higher frequencies.

In this chapter, numerical SAR assessment is performed on human head-equivalent cube models (HHECM) and human eye-equivalent cube models (HEECM) between 2 and 30 GHz in order to find out how SAR can be used as useful in limiting the over exposure at mmWs. For this purpose, EM wave penetration depth is used to identify the absorption levels with the help of SAR inside the body from the surface.

3.2 Exposure Configuration

There is a well-established and well-accepted thermal heating effect due to the mmW exposure. The primary targets at mmWs, organs, or biological tissues are the human eyes and skin. The human skin thickness varies from 0.5 mm on eyelids to 4 mm on the heels of feet [10, 12]. Skin is the largest tissue among the other human tissues and organ. It is well understood that tissues with high concentration of water absorb more energy. Therefore, eyes and skin are major concern at mmW frequency range.

For this purpose, single-layer (skin only), multi-layer (skin, fat and muscle) HHECM and HEECM were exposed with plane wave and SAR values were calculated. The IPD was set to 2.65 mWm^{-2} for all exposure scenarios. The exposure configuration of single- and multi-layer model is shown in Figure 3.1.

Dielectric properties of tissues such as dry- and wet-skin, fat, and muscle were calculated over the wide frequency range from 2 to 30 GHz using the Debye model. First order equation of Debye model is given as follows:

$$\varepsilon(\omega) = \varepsilon_\infty + \frac{\varepsilon_s - \varepsilon_\infty}{1 + i\omega\tau} + \frac{\sigma}{i\omega\varepsilon_0} \tag{3.1}$$

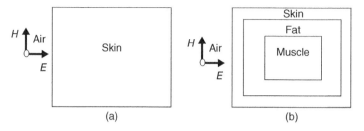

Figure 3.1 Plane wave exposure configuration of (a) Single-layer skin-equivalent cube model. (b) Multi-layer skin, fat and muscle-equivalent cube model.

All relaxation parameters of the Debye model are calculated based on curve fitting technique. The dielectric properties, namely, permittivity and conductivity graphs, are presented in Figures 3.A.1 and 3.A.2.

3.3 Plane Wave Exposure Assessment of E-field Absorption Within the Skin Using SAR as a Function of Frequency

In this assessment configuration, dry- and wet-skin tissue dielectric properties were assigned to a single-layer HHECM in two separate cases. The HHECM was designed in CST microwave studio and the dimensions were set to $120 \times 170 \times 180 \, \text{mm}^3$ as sown in Figure 3.2.

The model was exposed with plane wave (as shown in Figure 3.1a) in free space at various frequencies (2, 6, 10, 12, 16, 20, 24, 28, and 30 GHz), where the E- and H-field components were aligned perpendicular and parallel to the cube model, respectively. Standard boundary conditions were considered. The purpose of choosing cube volume was to investigate the EM absorption levels inside the model

Figure 3.2 3D view of HHECM, where all dimensions are in mm.

with respect to its surface. The cube SAR was numerically calculated and iterative computation on its central sub-volume was applied to calculate SAR values to assess the depth at which the wave have no longer effect on SAR performance.

3.3.1 Comparisons of SAR Levels on Dry-Skin and Wet-Skin

After exposing the HHECM with plane wave resultant SAR values are compared in both dry- and wet-skin cases at previously mentioned frequencies. Initially SAR was numerically calculated in whole cube model then SAR values were also calculated in its sub-volume (as shown in Figure 3.3) in order to investigate the EM waves penetration depth.

SAR values were calculated in sub-volume after reducing the HHECM size by setting "d" 1, 2.5, 5, 10, 20, and 50 mm from the surface of model from all sides. The size reduces from all sides in order to keep the consistency of thickness of human skin. Similar reduction method applied to rest five cases.

It is quite evident from Figure 3.4 that there is clear difference among the EM absorption levels within the dry- and wet-skin cases. The whole body specific absorption rate (WBSAR) values of HHECM (no-reduction) and its sub-volume cases (when size reduced by considering "d" is 1, 2.5, and 5 mm) are almost similar in both dry- and wet-skin cases, which is due to the deeper penetration of EM waves at 2 GHz. It means that EMFs absorbs not only on the surface of skin but deep inside the body. Approx. 57 and 50% energy absorbs within 20 mm of skin from the surface in dry- and wet cases correspondingly.

Another important fact is less absorption noticed in wet-skin as compared to dry-skin exposure scenarios. It is usually noticed that when wet-skin exposed to EM waves, it absorbs more energy but the dry-skin has shown higher absorption levels at all frequency bands that are considered in this study. The possible reason behind this trend could be due to the very minor difference in dielectric properties.

EM waves penetration depth decrease quickly as frequency gradually increases as can be seen in Figures 3.5–3.7. The difference between the WBSAR values of

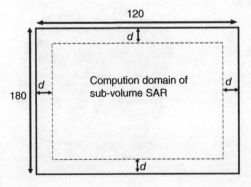

Figure 3.3 Sub-volume dimensions of HHECM; where "d" is the variable distance from the cube surface, where all dimensions are in mm.

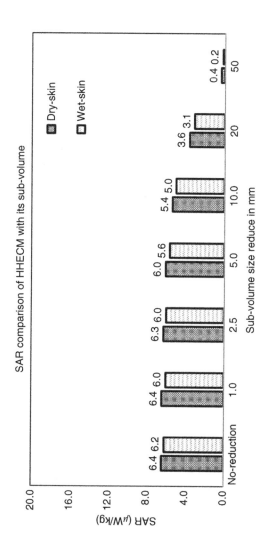

Figure 3.4 Comparison of SAR values of dry- and wet-skin HHECM with its sub-volume resultant from plane wave exposure at **2 GHz**, where no-reduction key word on x-axis of graph represents the SAR in whole cube volume. It can be seen that the wave at this frequency doesn't exceed 50 mm from the HHECM surface.

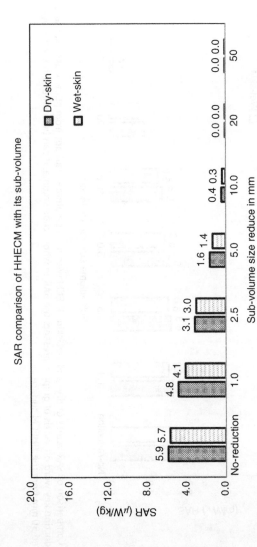

Figure 3.5 Comparison of SAR values of dry- and wet-skin HHECM with its sub-volume resultant from plane wave exposure at **6 GHz**, where no-reduction key word on x-axis of graph represents the SAR in whole cube volume. It can be seen that the wave at this frequency doesn't exceed 10 mm from the HHECM surface.

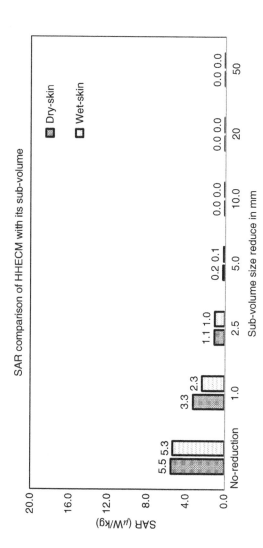

Figure 3.6 Comparison of SAR values of dry- and wet-skin HHECM with its sub-volume resultant from plane wave exposure at **10 GHz**, where no-reduction key word on x-axis of graph represents the SAR in whole cube volume. It can be seen that the wave at this frequency doesn't exceed 5 mm from the HHECM surface.

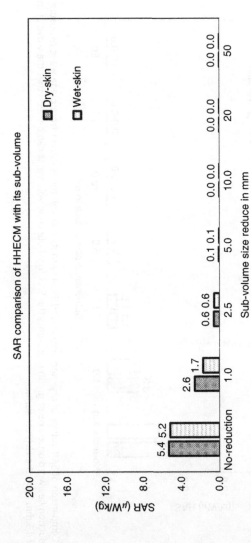

Figure 3.7 Comparison of SAR values of dry- and wet-skin HHECM with its sub-volume resultant from plane wave exposure at **12 GHz**, where no-reduction key word on x-axis of graph represents the SAR in whole cube volume. It can be seen that the wave at this frequency doesn't exceed 5 mm from the HHECM surface.

HHECM (no-reduction case) at 6 GHz with its sub volume cases when box phantom size reduced by considering "d" is 1 and 2.5 mm, which are within 20 and 48%, respectively (as shown in Figure 3.5). Almost 50% energy absorbed within the 2.5 mm of skin from surface and rest energy absorbs within 2.5–5 mm skin. The energy absorption within skin is negligible approaching to almost zero after 5 mm distance from skin surface to deep inside the HHECM.

At 10 and 12 GHz, 80 and 90% of energy is absorbed within the 2.5 mm skin from surface, respectively (as shown in Figures 3.6 and 3.7). The remaining 20 and 10 % energy absorbs between 2.5 and 5 mm skin at both frequency bands correspondingly. On the other hand, energy absorption within the wet-skin shows the similar behavior in comparison with dry-skin case.

E-field values calculated based on the resulted SAR values in both dry- and wet-skin scenarios as shown in Figures 3.8 and 3.9, respectively. It can be clearly seen that E-field absorption levels are much higher and its penetration depth is deeper at lower frequencies (2 GHz). E-field absorption level and its penetration depth gradually decreases with the increase in frequency, which is approx. zero at 10 mm distance from the surface of skin at 12 GHz in both dry- and wet-skin cases (as shown in Figures 3.8 and 3.9).

The graph presented in Figure 3.16 represents the penetration of EM waves within the HHECM with respect to frequency. It can be clearly seen that EM absorption within the dry- and wet-skin equivalent cube model doesn't exceed 5 at

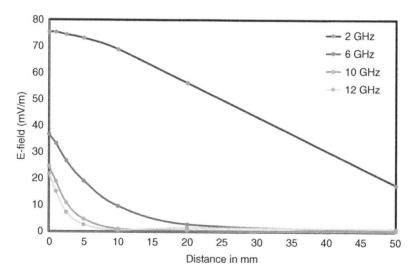

Figure 3.8 E-field absorption within the dry-skin from surface to 50 mm deep inside the HHECM.

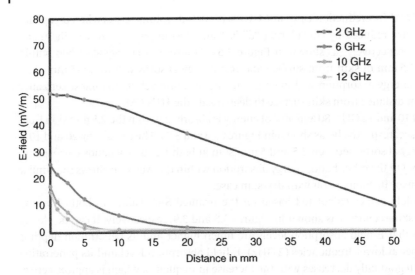

Figure 3.9 E-field absorption within the wet-skin from surface to 50 mm deep inside the HHECM.

12 GHz. Further simulations were conducted to investigate EM absorption levels on dry-skin only above 12 GHz. It was found that almost 100% of energy absorbs within 1.5 mm of skin from its surface above 16 GHz. All the SAR values resultant from plane wave exposure at 12–30 GHz are presented in Figures 3.A.3–3.A.7.

3.4 Plane Wave Exposure Assessment of E-field Absorption Within Multi-layer Model Using SAR as a Function of Frequency

In this assessment configuration, three-layer dry-skin, fat, and muscle-equivalent HHECM were exposed by plane wave source (as shown in Figure 3.1), where the E- and H-field components were aligned perpendicular and parallel to the cube model, respectively. Standard boundary conditions were applied. All tissue properties were produced using Debye first order relaxation model. The choice of dry-skin instead wet-skin in multi-layered model was based on higher absorption levels observed when exposed with plane wave in single-layer HHECM.

The multi-layer model has same dimensions as single-layer model as shown in Figure 3.2. In this multi-layer configuration skin and fat thickness was set 4 and 25 mm, respectively, and remaining cube was considered as muscle. The choice of thickness of skin, fat, and muscle is based on data available in literature [13, 14]. The multi-layer model is composed of three cubes. The outer most cube is skin,

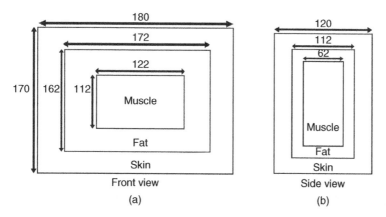

Figure 3.10 Dimensions of multi-layer HHECM; The outer, middle and inner cube represents skin, fat, and muscle, respectively, where all dimensions are in mm. (a) Front view. (b) Side view.

middle cube is fat, and inner cube is muscle. The composition of three-layer model is shown in Figure 3.10.

3.4.1 Comparisons of SAR Levels on Dry-Skin and Multi-layer Model

After exposing the HHECM with plane wave resultant, SAR values are compared in both dry-skin and multi-layered models at previously mentioned frequencies. Initially SAR was numerically calculated in whole cube model then SAR values were also calculated in its sub-volume (as shown in Figure 3.3) in order to investigate the EM waves penetration depth. The SAR calculation method is exactly the same as performed in single-layer model. Due to the huge number of mesh cells generated in this multi-layer configuration, simulations were only conducted up to 12 GHz.

SAR values were also calculated in its sub-volume same as single-layer exposure configuration, and sub-volume size achieved by reducing the HHECM size by considering "d" is 1, 2.5, 5, 10, 20, and 50 mm from the surface of model from all sides. The size reduces from all sides in order to keep the consistency of thickness of human skin. Similar reduction method applied to rest five cases. Higher SAR values are observed in dry-skin as compared to multi-layer model as shown in Figure 3.11.

It is understandable that single-layer skin equivalent exposure cases has shown slightly over estimated SAR values as compared to multi-layer cases. But interestingly multi-layer model has shown higher absorption levels in 20 and 50 mm reduction cases. The EMFs absorption level at 20 mm distance from surface of skin is approx. 77 and 57% in multi-layer model and dry-skin, respectively. It is

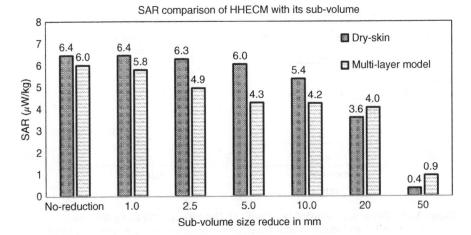

Figure 3.11 Comparison of SAR values of dry-skin and multi-layer HHECM with its sub-volume resultant from plane wave exposure at **2 GHz**, where no-reduction key word on *x*-axis of graph represents the SAR in whole cube volume. It can be seen that the wave at this frequency doesn't exceed 50 mm from the HHECM surface.

worth mentioning theoretically penetration depth of EM waves in multi-layer model is higher. Therefore, absorption levels are significantly higher as compared to dry-skin at 2 GHz.

EM absorption levels are significantly higher in multi-layer model and in its sub-volume as compared to dry-skin cases at 6 GHz (as shown in Figure 3.12). Approx. 10–20% higher SAR values noticed in multi-layer model and in its sub-volume in comparison with dry-skin. In multi-layer model approx. 50 and 20% of energy absorbs within 2.5 and 10 mm cases from its surface correspondingly.

At 10 and 12 GHz, approx. 57 and 48% energy absorbs within 1 mm from the surface of multi-layer model, respectively (as shown in Figures 3.13 and 3.14). It can be clearly seen that at both frequencies more than 90% of energy absorbs within 2.5 mm of multilayer model and similar behavior noticed in dry-skin. It is worth mentioning that above 2 GHz higher SAR values found in multi-layer model as compared to single-layer model.

This could be the reason why international guidelines providing organization such as FCC/IEEE/ICNIRP emphasis to use dielectric properties for homogenous model that can provide slightly over estimated SAR values as compared to heterogeneous model. Moreover, this finding is in line with the reference [6, 8, 9], where higher level of temperature variation observed in multi-layer models as compared to single-layer model at 60 GHz. So it can be concluded that as you go higher in the frequency homogenous models may not be suitable for SAR assessments or in other words for temperature variations. Multilayer models might be the only

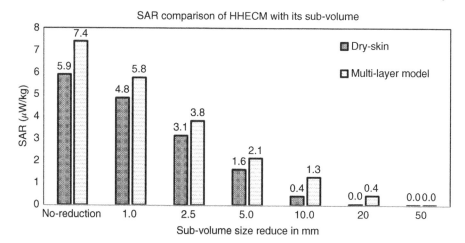

Figure 3.12 Comparison of SAR values of dry-skin and multi-layer HHECM with its sub-volume resultant from plane wave exposure at **6 GHz,** where no-reduction key word on *x*-axis of graph represents the SAR in whole cube volume. It can be seen that the wave at this frequency doesn't exceed 20 mm from the HHECM surface.

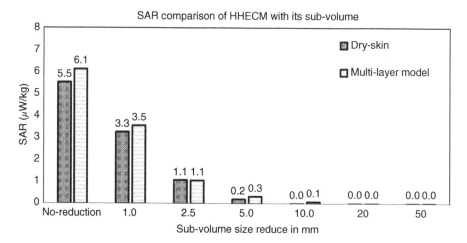

Figure 3.13 Comparison of SAR values of dry-skin and multi-layer box phantom with its sub-volume resultant from plane wave exposure at **10 GHz,** where no-reduction key word on *x*-axis of graph represents the SAR in whole cube volume. It can be seen that the wave at this frequency doesn't exceed 10 mm from the HHECM surface.

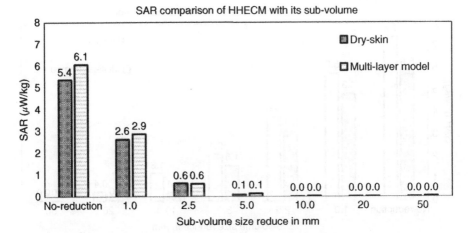

Figure 3.14 Comparison of SAR values of dry-skin and multi-layer box phantom with its sub-volume resultant from plane wave exposure at **12 GHz**, where no-reduction key word on x-axis of graph represents the SAR in whole cube volume. It can be seen that the wave at this frequency doesn't exceed 5 mm from the HHECM surface.

option left to consider to accurately assess the EM absorption levels either using by SAR or temperature variation.

E-field values calculated based on the resulted SAR values in multi-layer model as shown in Figure 3.15. It can be clearly seen that E-field absorption levels are

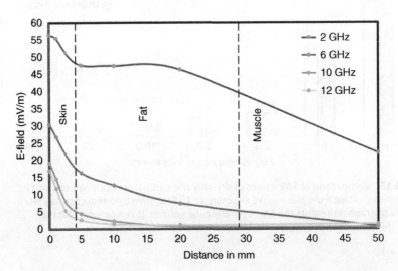

Figure 3.15 E-field absorption within the multi-layer model (dry-skin, fat, muscle) from surface to 50 mm deep inside the HHECM.

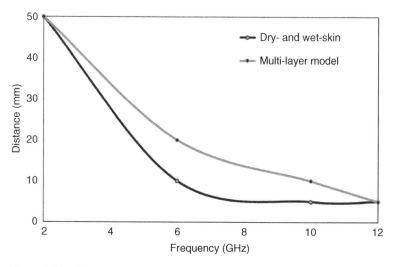

Figure 3.16 Comparison of maximum penetration depth of EM waves within dry- and wet-skin equivalent HHECM and multi-layer HHECM from its surface.

much higher and its penetration depth is deeper at lower frequencies (2 GHz). E-field absorption levels reaches to 0 mV/m at 20 mm distance from the surface of HHECM at 12 GHz. The graph presented in Figure 3.16 represents the penetration of EM waves within the HHECM with respect to frequency. It can be clearly seen that EM absorption within multi-layer HHECM is much higher between 2 and 10 GHz as compared to dry- and wet-skin equivalent cube model. But interestingly EM absorption levels in both exposure configurations doesn't exceed 5 mm at 12 GHz.

3.5 Plane Wave Exposure Assessment of E-field Absorption Within the Eye Using SAR as a Function of Frequency

In this assessment configuration, HEECM was exposed with plane wave source in free space, where the E- and H-field components were aligned perpendicular and parallel to the cube model, respectively. The HEECM dimensions was set to 24 mm³ [15, 16] and standard boundary condition was applied. The cube model is composed of two tissue layers i.e. outer and inner layers. The outer layer thickness was set to 1 mm³ and cornea tissue properties were assigned to it. Similarly inner layer thickness was set to 23 mm³ and vitreous humor tissue properties were assigned to it (as shown in Figure 3.17). The electrical properties of both tissues were calculated using the Debye first-order relaxation model.

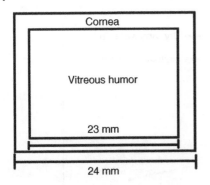

Figure 3.17 HEECM composition with its dimensions.

3.5.1 Comparisons of SAR Levels on HEECM and Multi-layer Model

The SAR values are calculated in both HEECM and multi-layered model (size normalized to HEECM equivalent) between 10 and 30 GHz. Initially SAR was numerically calculated in whole cube model then SAR values were also calculated in its sub-volume (as shown in Figure 3.3) in order to investigate the EM waves penetration depth. The SAR calculation method is exactly the same as performed in previous sections. Simulations were only conducted up to 30 GHz due to fact that penetration depth of EM waves is within 2.5 mm of distance from its surface in the case of HEECM. The SAR results at 30 GHz are discussed in next section and results from 10 to 28 GHz are presented in Figures 3.A.8–3.A.13. Due to fact that the same phenomenon of penetration depth noticed is discussed in previous sections.

SAR values were also calculated in HEECM and multi-layer model (HEECM equivalent) sub-volumes same as single-layer and multi-layer exposure configurations. At 30 GHz, it can be clearly seen that approx. 95 and 90% of energy absorbs with 1 mm from the surface of HEECM and multi-layer model, respectively (as shown in Figure 3.18). It was initially assumed that HEECM will show the higher SAR values as compared to multi-layer model due to the fact that eye tissues have high concentration of water. Interestingly multi-layer model shows higher SAR values as compared to HEECM. There were some higher SAR values found deep inside the multi-layer model at 5 and 10 mm distance from its surface. This is due to the fact that at 30 GHz, a huge number of mesh cells are required to improve the accuracy of SAR estimation.

In addition, maximum SAR values were also calculated in both HEECM and multi-layer models between 10 and 30 GHz frequency ranges (as shown in Figure 3.19). Due to the shallow penetration depth of EM waves and small size of human eye (human eye weights 14 g), maximum SAR values were averaged over 0.1 g cube instead of 1 or 10 g cube. There are approx. 33 and 10% higher maximum SAR values found in the eye and multi-layer models, respectively, at

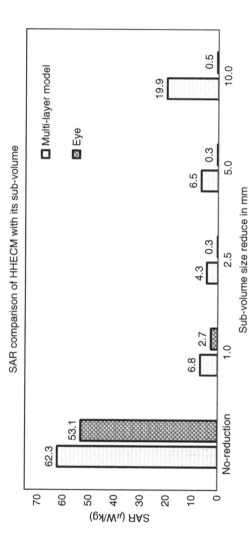

Figure 3.18 Comparison of SAR values of HEECM and multi-layer model (size normalized to HEECM equivalent) with its sub-volume resultant from plane wave exposure at **30 GHz**, where no-reduction key word on *x*-axis of graph represents the SAR in whole cube volume. It can be seen that the wave at this frequency doesn't exceed 2 mm or less from the HEECM surface.

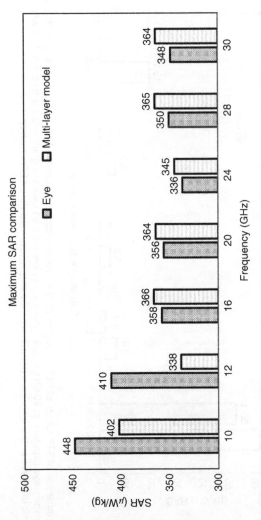

Figure 3.19 Comparison of maximum SAR values between HEECM and multi-layer model (size normalized to HEECM equivalent) resultant from plane wave exposure. The maximum SAR values calculated in 0.1 g cube.

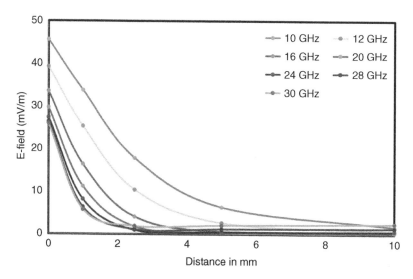

Figure 3.20 E-field absorption within the eye from surface to 50 mm deep inside the HEECM.

10 GHz as compared to 30 GHz. It is worth mentioning that maximum SAR values between 16 to 30 GHz found 3–5% higher in multi-layer model in comparison with the eye.

E-field values calculated based on the resulted SAR values in eye as shown in Figure 3.20. It can be clearly seen that E-field absorption levels are much higher and its penetration depth is deeper at lower frequencies (10 GHz). E-field absorption level gradually decreases with the increase in frequency, which is approx. 2 mV/m at 10 mm distance from the surface of HEECM at 30 GHz. The graph presented in Figure 3.21 represents the penetration of EM waves within the HEECM with respect to frequency. It can be clearly seen that between 16 GHz to 30 GHz the penetration depth EM waves is within 2–3 mm from the surface of HEECM.

However, it has been previously mentioned that present regulations at mmWs provide safety limits in term of IPD. But without providing any specific methodologies and evaluations guidelines for exposure under near-field configurations. In this chapter, a method is provided to calculate the SAR within few millimeter of biological material. Recently an empirical method is provided in reference [17] to calculate IPD using the SAR and penetration depth. The equation for IPD calculation is given in the following [17]:

$$\text{IPD}[\text{W/m}^2] = \frac{\rho \ \delta \ \text{SAR}(0)}{2 \ (1 - |\Gamma|^2)} \tag{3.2}$$

where "ρ" is the mass density, "δ" is the penetration depth, SAR (0) is the SAR at the skin surface, and "Γ" is the reflection coefficient.

The numerical technique is presented in this chapter to calculate SAR (within few millimeter of any biological material) and penetration depth, which can be

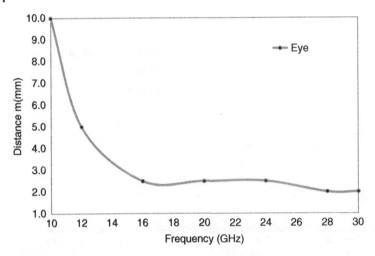

Figure 3.21 Maximum penetration depth of EM waves within eye from the surface of HEECM.

very useful in predicting IPD levels with some accuracy. It is worth mentioning that SAR assessment at the surface of any biological material is extremely difficult and challenging due to the cube averaging method.

3.6 Chapter Summary

This chapter presents the review on current understanding of the biological effects caused by exposure from mmW frequencies. International guidelines providing agencies recommend to use PD as a basic restriction threshold to limit the human body excessive exposure above 6 and 10 GHz for Federal Communications Commission Office (FCC) and International Commission on Non-Ionizing Radiation Protection (ICNIRP), respectively. The difficulty is how SAR can be applicable when average volume is different and also due to the shallow penetration depth at mmWs. For the purpose of understanding SAR assessment is performed numerically from 2 to 30 GHz in single- and multi-layer HHECM, and HEECM in order to investigate penetration depth using SAR and the absorption levels within few millimeter from the surface of exposed biological material.

HHECM and HEECM were exposed with plane wave and IPD was set to 2.65 mWm^{-2} in all exposure scenarios. SAR values are also calculated in its sub-volume domain to investigate the depth at which the EM waves have no longer effect on SAR estimation in both homogenous and heterogeneous cases. It is concluded that approximately 90% of energy absorbs within 1 mm of skin above 30 GHz in homogenous cases. It is worth mentioning that multi-layer model (size normalized to HEECM) has shown higher SAR values compared to HEECM above 16 GHz.

Appendix 3.A

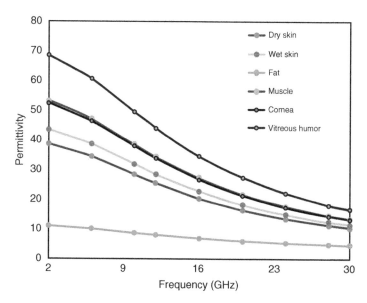

Figure 3.A.1 Debye model fitted permittivity values of dry- and wet-skin, fat, muscle, cornea, vitreous humor.

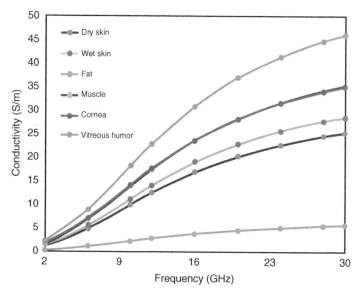

Figure 3.A.2 Debye model fitted conductivity values of dry- and wet-skin, fat, muscle.

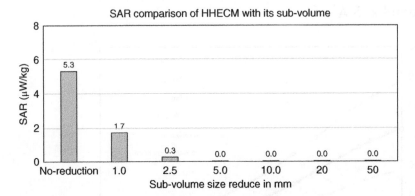

Figure 3.A.3 Comparison of SAR values of dry-skin equivalent HHECM with its sub-volume resultant from plane wave exposure at **16 GHz**.

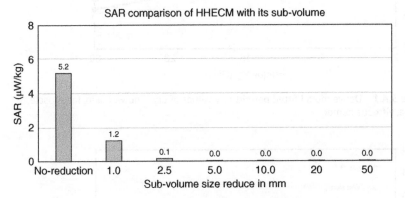

Figure 3.A.4 Comparison of SAR values of dry-skin equivalent HHECM with its sub-volume resultant from plane wave exposure at **20 GHz**.

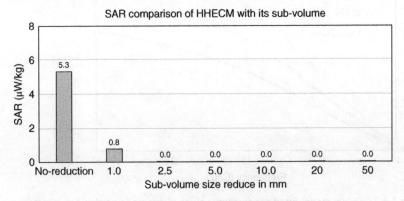

Figure 3.A.5 Comparison of SAR values of dry-skin equivalent HHECM with its sub-volume resultant from plane wave exposure at **24 GHz**.

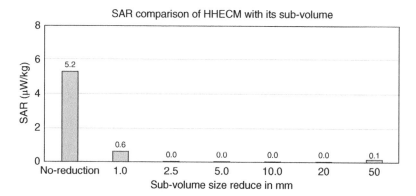

Figure 3.A.6 Comparison of SAR values of dry-skin equivalent HHECM with its sub-volume resultant from plane wave exposure at **28 GHz**.

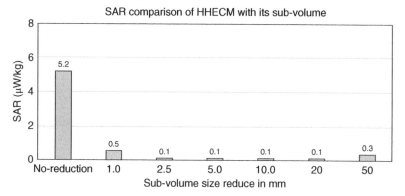

Figure 3.A.7 Comparison of SAR values of dry-skin HHECM with its sub-volume resultant from plane wave exposure at **30 GHz**.

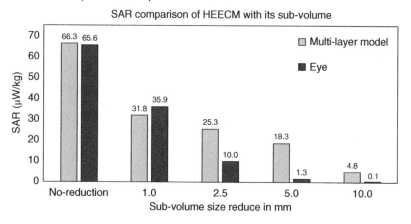

Figure 3.A.8 Comparison of SAR values of HEECM and multi-layer model (size normalised to HEECM equivalent) with its sub-volume resultant from plane wave exposure at **10 GHz**, where no-reduction key word on *x*-axis of graph represents the SAR in whole cube volume. It can be seen that the wave at this frequency doesn't exceed 5 mm from the HEECM surface.

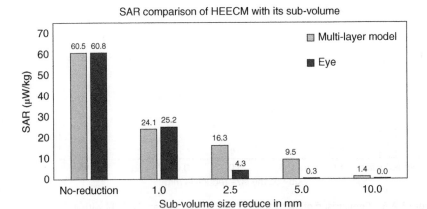

Figure 3.A.9 Comparison of SAR values of HEECM and multi-layer model (size normalised to HEECM equivalent) with its sub-volume resultant from plane wave exposure at **12 GHz**, where no-reduction key word on *x*-axis of graph represents the SAR in whole cube volume. It can be seen that the wave at this frequency doesn't exceed 5 mm from the HEECM surface.

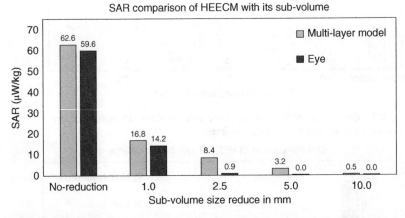

Figure 3.A.10 Comparison of SAR values of HEECM and multi-layer model (size normalised to HEECM equivalent) with its sub-volume resultant from plane wave exposure at **16 GHz**, where no-reduction key word on *x*-axis of graph represents the SAR in whole cube volume. It can be seen that the wave at this frequency doesn't exceed 2.5 mm from the HEECM surface.

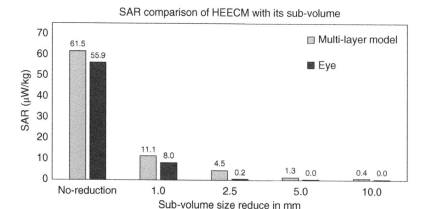

Figure 3.A.11 Comparison of SAR values of HEECM and multi-layer model (size normalised to HEECM equivalent) with its sub-volume resultant from plane wave exposure at **20 GHz**, where no-reduction key word on *x*-axis of graph represents the SAR in whole cube volume. It can be seen that the wave at this frequency doesn't exceed 2.5 mm from the HEECM surface.

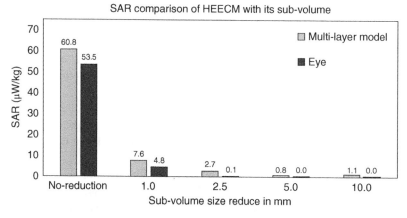

Figure 3.A.12 Comparison of SAR values of HEECM and multi-layer model (size normalised to HEECM equivalent) with its sub-volume resultant from plane wave exposure at **24 GHz**, where no-reduction key word on *x*-axis of graph represents the SAR in whole cube volume. It can be seen that the wave at this frequency doesn't exceed 2.5 mm from the HEECM surface.

Figure 3.A.13 Comparison of SAR values of HEECM and multi-layer model (size normalised to HEECM equivalent) with its sub-volume resultant from plane wave exposure at **28 GHz**, where no-reduction key word on *x*-axis of graph represents the SAR in whole cube volume. It can be seen that the wave at this frequency doesn't exceed 2.5 mm from the HEECM surface.

References

1 ICNIRP (1998). International Commission on Non-Ionizing Radiation Protection: guidelines for limiting exposure to time-varying electric, magnetic and electromagnetic fields (up to 300 GHz). *Health Physics Society* 74: 494–522.

2 Cleveland, R.F., Sylvar, D.M., and Ulcek, J.L. (2001). *Evaluating Compliance with FCC Guidlines for Human Exposure to Radiofrequency Electromagnetic Fields*, 1–52. Washington DC: Federal Communication Commission Office of Engineering & Technology.

3 Rappaport, T.S., Sun, S., Mayzus, R. et al. (2013). Millimeter wave mobile communications for 5G cellular: it will work! *IEEE Access* 1: 335–349.

4 Alekseev, S.I., Ziskin, M.C., and Fesenko, E.E. (2011). Problems of using a thermocouple for measurements of skin temperature rise during the exposure to millimeter waves. *Biophysics* 56: 561–565.

5 Alon, L., Sloovinsky, W.S., Cho, G.Y., and Rappaport, T.S. (2015). mmWaves Exposure assessment using magnetic resonance thermal imaging. Presented at the BioEM, Monterey conference. BioEM2015, Monterey, USA, June 14–19, 2015.

6 Kanezaki, A., Watanabe, S., Hirata, A., and Shirai, H. (2008). Theoretical analysis for temperature elevation of human body due to millimeter wave exposure. International Biomedical Engineering Conference in Cairo, pp. 1–4.

7 Kanezaki, A., Hirata, A., Watanabe, S., and Shirai, H. (2009). Effects of dielectric permittivities on skin heating due to millimeter wave exposure. *Biomedical Engineering* 8: 1–9.

8 Kanezaki, A., Hirata, A., Wantanabe, S., and Shirai, H. (2010). Parameter variation effects on temperature elevation in a steady-state, one-dimensional thermal model for millimeter wave exposure of one- and three-layer human tissue. *Physics in Medicine and Biology* 55: 4647–4659.

9 Wu, T., Rappaport, T.S., and Collins, C.M. (2015). Safe for generations to come: considerations of safety for millimeter waves in wireless communications. *IEEE Microwave Magazine* 16: 65–84.

10 Guraliuc, A., Zhadobov, M., and Sauleau, R. (2014). Beyond 2020 heterogeneous wireless network with millimeter-wave small-cell access and backhauling. MiWaveS.

11 Zhadobov, M., Chahat, N., Sauleau, R. et al. (2011). Millimeter-wave interactions with the human body: State of knowledge and recent advances. *International Journal of Microwave and Wireless Technologies* 3: 237–247.

12 Ha, R.Y., Nojima, K., Adams, W.P., and Brown, S.A. (2005). Analysis of facial skin thickness: defining the relative thickness index. *Plastic and Reconstruction Surgery* 115: 1769–1773.

13 Lee, S.H., Kim, D., Baek, M.Y. et al. (2015). Abdominal subcutaneous fat thickness measured by ultrasonography correlates with hyperlipidemia and steatohepatitis in obese children. *Pediatric Gastroenterology, Hepatology & Nutrition* 18: 108–114.

14 Black, D., Vora, J., Hayward, M., and Marks, R. (1988). Measurement of subcutaneous fat thickness with high frequency pulsed ultrasound: comparisons with a caliper and a radiographic technique. *Clinical Physics and Physiological Measurement* 9: 57–64.

15 Diameter of a human eye. https://hypertextbook.com/facts/2002/AniciaNdabahaliye1.shtml (accessed August 2017).

16 Human eye. https://application.wiley-vch.de/books/sample/3527403809_c01.pdf (accessed August 2017).

17 Guraliuc, A.R., Zhadobov, M., Sauleau, R. et al. (2017). Near-field user exposure in forthcoming 5G scenarios in the 60-GHz band. *IEEE Transactions on Antennas and Propagation* 65 (12): 6606–6615.

4

Age Dependent Exposure Estimation Using Numerical Methods

Muhammad Rafaqat Ali Qureshi[1], Yasir Alfadhl[1], Xiaodong Chen[1], and Masood Ur Rehman[2]

[1] School of Electronic Engineering and Computer Science, Queen Mary University of London, London, UK
[2] James Watt School of Engineering, University of Glasgow, Glasgow, UK

4.1 Introduction

The concept of controlling homes appliances through wireless links makes modern smart homes. These homes and exponential increase in use of wireless devices has raised concern over excessive exposure from these devices. Experimentally it's very difficult to realize different realistic exposure configurations or estimate EM absorption within the human body. For this purpose, anatomical human models of various age groups used in numerical dosimetry.

In this chapter, numerical anatomical human models representing a sample of the population have been configured and modelled using an established commercially available computation technique, namely, CST Microwave Studio (MWS) [1]. These models have been simulated with very fine Cartesian volume cells (voxels) resolutions of up to 2 mm to ensure adequate representation of each organ, and to satisfy the numerical stability requirements. The dielectric properties for all models were calculated based on Cole-Cole formulations [2], and the Nello Carrara Institute of Applied Physics (IFAC) online calculator [3], which is based on data reported in references [4–6]. Where no data were available for a particular tissue, adjustments and replacements were made to assign values from tissues with similar compositions.

Furthermore, age-dependent effects on dielectric properties were realized by utilizing the novel age- and frequency-dependent (AFD) method. In this method, three different techniques are developed to calculate AFD dielectric properties.

Relaxation parameters for Cole–Cole model were also produced and compared with different source and good agreement achieved to validate this method. This method utilizes the measured data of animals (porcine/pigs) tissues using a curve-fitting techniques to map the AFD dielectric properties to their appropriate human equivalent ones [2, 7].

CST MWS widely used in the research community and has been further validated and used in this study. Details of the validation studies as well as the modelling and selection of dielectric properties for all age groups are discussed in detail in this chapter.

4.2 Numerical Human Models

In this section modelling of voxel models is discussed in detail. The voxel models represent the complex structure of organs. Each organ has its own electrical properties and it absorbs different levels of energy when exposed to EM waves. A wide range of voxel models are available in literature [8]. However, in this section modelling of three different voxel models i.e. two adult and one child model discussed in detail and later used to study the specific energy absorption rate (SAR) performance due to the exposure from wireless devices. The details about the voxel models are given in section.

4.2.1 Adult Voxel Models

Two adult voxel models (one female and one male) consider for EM evaluation. The female model is based on a voxelized 23-year-old female model, which is constructed from 791 slices along the body's long axis, with 294×124 voxels in the horizontal plane. By considering voxel dimensions of 1.9, 2.0, and 2.0 mm in the x, y, and z voxel axis, respectively.

The resulting model has a height of 1.63 m and a weigh around 65 kg. These parameters are very close to the work presented in [9], which produced 1.63 m and mass of 60 kg. The difference in weight is attributed to the voxel dimensions and the mass densities used. Figure 4.1 shows a representation of the female model, with the opacity of the voxels varied to display the various body organs and tissue types. There were 40 different tissue types modelled in female model. Each of the tissue-voxel types were populated with their corresponding dielectric parameters, calculated from IFAC database [3] as given in Table 4.A.1.

The adult 34-year-old male model (shown in Figure 4.2) is based on a voxelized male model, which is constructed from 871 slices along the body's long axis, with

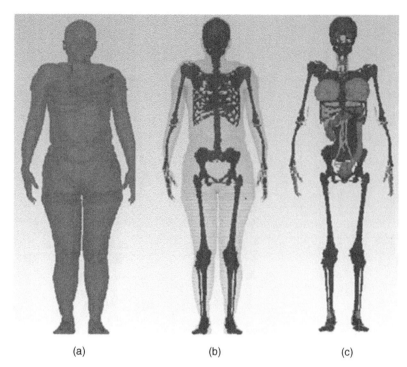

(a)   (b)   (c)

Figure 4.1 Volume rendered images of the female model; (a) The outside surface; (b) Skin and skeleton; (c) Skeleton with a few internal organs (skin, fat, and muscle removed). Rendering resolution is 2 mm³.

148×277 voxels in the horizontal plane. The computed male adult model represents a 1.76 m tall and weighs around 76 kg with voxel dimensions of 1.9, 2.0, and 2.0 mm in each of the voxel axis. These dimensions are very close to work presented in reference [9], where the adult male has a height of 1.76 m and mass of 73 kg. The difference in weight is attributed to the mass densities used as mentioned in previous section. The dielectric properties of the 38 different tissue types included in male model were also calculated from the online database [3] as given in Table 4.A.2.

4.2.2 Child Voxel Model

The 7-year-old female child model (shown in Figure 4.3) is based on the ITIS Eartha virtual human model [8], representing 0.965 m tall and weigh of around 21.7 kg. The voxel dimensions are of 1.4, 1.4, and 1.4 mm in x, y, and z axis,

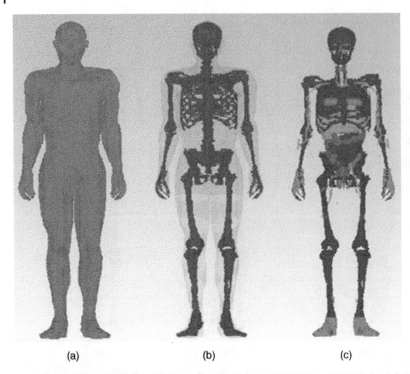

(a)	(b)	(c)

Figure 4.2 Volume rendered images of the male model; (a) The outside surface; (b) Skin and skeleton; (c) Skeleton with a few internal organs (skin, fat, and muscle removed); Rendering resolution is 2 mm^3.

respectively. There are 300×122 voxels in each horizontal plane. A few refinements were made to the dielectric properties of the tissues to ensure that they are valid for the given age and for the operational frequencies of SMs such as 868 and 2450 MHz. This female child model has 75 different tissue types that are given in Table 4.A.3. The dielectric properties of tissues used in each of the models have been determined from Cole–Cole models [2, 5, 7].

The dielectric properties for the child model were calculated using novel age-dependent methods that are calculated from 10, 50, and 250 kg pigs data [2]. Moreover, growth curve data for a 21.3 kg pig [10] were used to interpolate the dielectric properties for the child model, and the same method was applied for all tissues to get equivalent data for a 7-year-old child. In recent years, animals have been found to be similar to humans in terms of body organ development. According to reference [10], 4 week old pig is comparable to a 7-year-old child, and therefore data for a 21.3 kg pig was used to assign the dielectric properties for the child model.

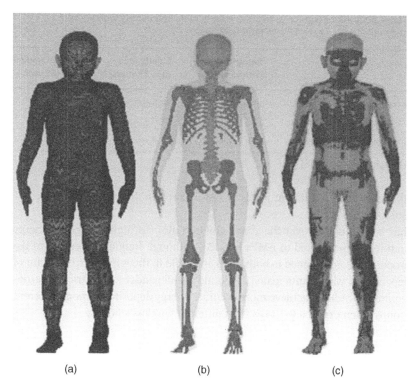

(a)   (b)   (c)

Figure 4.3 Volume rendered images of the female child model; (a) The outside surface; (b) Skin and skeleton; (c) Skeleton with a few internal organs (skin, fat, and muscle removed); Rendering resolution is 1.4 mm³.

4.3 Age-Dependent Tissue Properties

The frequency-dependent nature of biological tissues has been well studied over the decades. Frequency-dependent dielectric properties of a wide range of tissues have become widely available following publication of the Cole–Cole dispersion models [4–6]. Due to lack of availability of age-dependent dielectric properties, most studies generally consider the similar tissue properties for both adult and child models during the assessment of exposure from EMFs.

On the contrary, it is well-established that the water content level within the tissues of human body at different development stages (aging) becomes different. In literature, estimation methods based on empirical equations were used over a group of volunteers to show variations of total body water (TBW) as a function of age and weight as shown in Table 4.1 [11]. As a result of such differences in water contents, some studies have suggested that dielectric properties can be different,

Table 4.1 Body water (kg) for young and elderly men, and elderly women from published equations; where TBW is total body water.

	Young men (18–33 years)	Elderly men (67–89 years)	Elderly women (67–89 years)
TBW [measured]	40.4 ± 1.4	34.8 ± 0.7	34.8 ± 0.7
TBW (Moore) [age, weight]	42.3 ± 0.8	37.0 ± 0.5	28.5 ± 0.7
TBW (Pierson) [age, weight]	48.0 ± 1.4	41.4 ± 0.9	29.9 ± 1.1

Source: Reproduced from [11].

and this may lead to different levels of EM absorption due to the excessive EM exposure.

It has been reported over the year's children absorbs higher levels of energy than adults when exposed to EMFs [12, 13]. Although frequency-dependent tissue properties were assigned to both adult and child in the references but factor of age dependency wasn't anticipated. Accurate age-dependent dielectric properties are required in order to achieve more accurate energy deposition among different age groups when exposed to EMFs from different wireless sources.

4.3.1 Measured Tissue Properties

Human body organ growth and dielectric properties are comparable with animal's data. According to literature, data about the dielectric properties of biological tissues are gathered from mature animals. Few older studies stated the systematic changes in dielectric properties of brain tissues are due to the ageing effect [14, 15]. Some studies investigated the changes in dielectric properties of group of rat tissues including brain tissue as a function of age [9]. Several attempts were made in the past to demonstrate the similarities between human and rats on the basis of development of their organs [9, 16].

It is well-established that some tissues show variation in their dielectric properties as a function of age [2, 9, 17]. The water content level of the young age (child) tissue is higher than adults [2, 9, 11, 18]. A miniature animal between 6 and 8 months of age develops adult human sized organs and structure as reported in reference [19]. In addition, the growth of pig organs such as heart and cardiovascular system from birth to 4 months old age is analogous; in comparison with the growth of the same system in humans into the mid teen age. Similar comparisons were made to relate animal weight with human age as listed in Table 4.2.

It is therefore important to consider the effect of age-dependent properties (if any) on SAR assessment [12]. The importance of including age-relevant properties is also highlighted by the International Commission on Radiological

Table 4.2 Human age predictions using animals weight [2, 10, 18].

Animal weight (kg)	Human ages (year)	Assumed age in this work
12.5	5–6	5.5
21.3	7–8	7.5
30.5	9–10	9.5
40.5	11–12	11.5
51.5	13–14	13.5
65	15–16	15.5
80	17–18	17.5
95	19–20	19.5
110	21–22	21.5

Protection (ICRP) guidelines, which suggest the human organ mass changes as a function of age [20].

4.3.2 Age-dependent Human Dielectric Properties Extraction from Measured Data

Age-dependent dielectric properties of humans have been estimated based on the measured data for pigs of 10, 50, and 250 kg [2, 18]. The authors have supported their statement by running parallels between different developmental stages of animals and humans.

The 10, 50, and 250 kg animals were assumed to be their equivalent human 1–4 years old child, 11–13 years old teenage, and adult, respectively. Based on this approach, the measured dielectric data can be further utilized to predict the dielectric properties of a wider age groups.

To relate the animal weight with corresponding human age, pigs' growth rate [10] is considered and theory from references [2, 18] was applied. So on the basis of growth rate, in terms of organ growth, comparisons were made between human and animals weight as shown in Table 4.2. It was assumed that a 7 kg animal is equivalent to a 3-year-old child. Similarly, 12.5 kg animal would be comparable to 5–6-year-old humans.

4.3.3 Novel Calculation Methods of Age-dependent Dielectric Properties

Three different novel empirical techniques were developed to predict age-dependent dielectric properties. These techniques were also validated accurately against available literature data. Dielectric properties were calculated in the frequency range of 0.4–10 GHz using the Cole–Cole parameters from [2]. In total 14 tissue properties were estimated as a function of age, and tissues were the bone marrow, cornea, dura mater, fat, grey matter, intervertebral disc center, intervertebral disc, long bone, skin, skull, spinal cord, tongue, and white matter.

Data about the different developmental stages of animal was reported in [2]. In order to correlate this data to more precise equivalent human tissues, pig weights were used at three points to produce a curve with the best fit method. Animal brain growths were used from [10], to find equivalent developmental stage for humans. According to [2, 18], a 4-week-old pig could be a reasonable alternative to a 3-year-old child, in terms of organ growth and weight.

The curve fitting was applied in two distinguish ways in order to simplify the process of implementation, i.e. single frequency age-dependent method and dispersive age-dependent method. The single frequency age-dependent theorem involves the basic idea of curve fitting technique at single frequency and in order to implement on wide frequency range huge number of equations to solve for both permittivity and conductivity. Whereas the dispersive age-dependent method is a simplified and generalized function of curve fitting that reduces the number of equations for wide frequency range (0.4–10 GHz). The details about both methods are explained in next sections.

4.3.3.1 Single Frequency Age-Dependent Method

In this method, curve fitting technique was applied on three points representing dielectric properties of 10–250 kg animal. To test the proposed method, a single frequency (1 GHz) has been chosen as an example, to generate two equations for permittivity and conductivity, on the basis of best fit method as shown in Figures 4.4 and 4.5.

Permittivity and conductivity of grey matter at 1 GHz can be obtained through the following equations:

$$y = -4 \times 10^{-6} x^2 - 0.0347x + 14.662 \tag{4.1}$$

$$y = -8 \times 10^{-7} x^2 - 0.0006x + 0.2551 \tag{4.2}$$

where y is the permittivity at 1 GHz and x is weight of the pig. The weight of each pig corresponds to an equivalent human age. Equation (4.1) yield values of 14.23 and 10.80 for permittivity of fat at 1 GHz relating to a 12.5 (5.5-year-old human) and 110 kg pig (21.5-year-old human), respectively. The difference in the permittivity values of fat tissue among the 5.5- and 21.5-year-old human is approx. 24%.

The conductivity values resultant from Eq. (4.2) are 0.25 and 0.18 Sm^{-1} of fat at 1 GHz relating to a 21.3 (5.5-year-old human) and 110 kg pig (21.5-year-old human), correspondingly. The difference in the permittivity values of fat tissue among the 5.5- and 21.5-year-old human is approx. 28%.

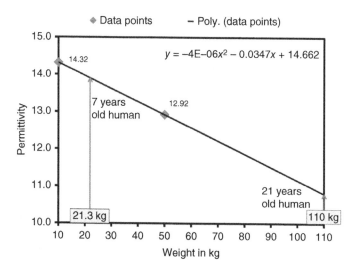

Figure 4.4 Permittivity of Fat at **1 GHz**.

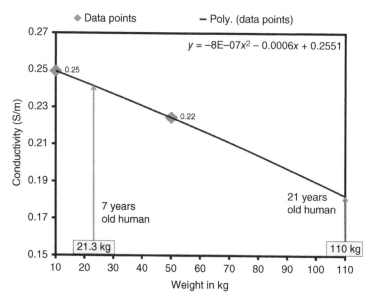

Figure 4.5 Conductivity of Fat at **1 GHz**.

Curve fitting technique was further applied to the available animal data in order to generate complex mathematical equations for each tissue type and then human age equivalent animal weight applied to produce the dielectric properties of desire ages.

The current method is applicable to single frequency only. But in order to calculate tissue properties at wide frequency range, a huge number of equations need to be produced for both permittivity and conductivity values. Therefore, an alternative method was introduced, which reduces the complexity of this method. The simplified method, known as dispersive age-dependent method, provides the generalized equation for all tissues except bone marrow. The details about new method are explained in Section 4.3.3.2.

4.3.3.2 Dispersive Age-Dependent Method

The dielectric behavior of both permittivity and conductivity of most tissues of animal can be described with the same empirical formula. The measured data about the bone marrow in [2] is based on 30 and 50% bone marrow. It is stated that the bone marrow was measured at two distances from the end of the bone and the distribution of bone marrow type changes as a function of animal age. Therefore, its behaviors can't be described with same empirical equation as for other tissues.

It was noted that the permittivity value decreases as pig weight increases. On the contrary, conductivity follows the opposite trend. The generalized empirical formula for AFD dielectric properties defined as given in the following text (Eq. 4.3), which can be used to verify the permittivity and conductivity of tissues within 0.4 to 10 GHz frequency range.

$$Y = A - Bx + C \tag{4.3}$$

The dielectric properties of tissues can be calculated by putting the parameters from Table 4.3 in the previous equation. The parameter "x" is the weight of pig equivalent to human age. The table also includes three sets of data for a particular tissue, which belongs from different frequencies.

It's worth mentioning that available online sources data corresponds to what human age is still unknown. An attempt is made by comparing animal data from reference [2] with available online sources [3, 21] at 2450 MHz (frequency chosen as an example) as shown in Table 4.4. It can be clearly seen in weight column that the dielectric properties from online sources doesn't belongs from same age groups.

To generate tissue properties for a wide frequency range, three data sets need to be calculates at particular age in the first step. Then, resultant three data points of permittivity and conductivity from Eq. (4.3) can be further fitted or joined separately to generate the dielectric properties at any frequency point as shown in Figure 4.6. The error associated with this method as compared to single frequency age-dependent method is approx. 1 and 10% for permittivity and conductivity, respectively.

The dielectric properties of fat have been calculated, as an example, at three different age groups (5.5, 13.5, and 21.5 years) over wide frequency range

Table 4.3 Parameters for age-dependent generalized expression (ID: Intervertebral Disc, IDC: Intervertebral Disc Centre).

Tissue	Frequency (GHz)	Permittivity			Conductivity		
		A	**B**	**C**	**A**	**B**	**C**
Cornea	0.4	$0.3x^2 \times 10^{-3}$	94.1×10^{-3}	57.61	$5 \times 10^{-6} x^2$	1.3×10^{-3}	0.95
	5	$0.3x^2 \times 10^{-3}$	98.1×10^{-3}	49.46	$2 \times 10^{-5} x^2$	7.2×10^{-3}	4.52
	10	$0.3x^2 \times 10^{-3}$	84.2×10^{-3}	41.88	$9 \times 10^{-5} x^2$	23.7×10^{-3}	10.65
Dura	0.4	$0.8x^2 \times 10^{-3}$	0.13	57.72	$4 \times 10^{-6} x^2$	2.3×10^{-3}	1.02
	5	$0.3x^2 \times 10^{-3}$	0.13	49.41	$2 \times 10^{-5} x^2$	6.2×10^{-3}	4.34
	10	$0.3x^2 \times 10^{-3}$	0.12	42.56	$5 \times 10^{-5} x^2$	17×10^{-3}	9.90
Fat	0.4	$-10^{-5} x^2$	33.9×10^{-3}	14.88	$-6 \times 10^{-7} x^2$	0.5×10^{-3}	0.21
	5	$4 \times 10^{-6} x^2$	31.1×10^{-3}	12.87	$3 \times 10^{-6} x^2$	4.3×10^{-3}	1.15
	10	$-10^{-5} x^2$	20.7×10^{-3}	10.68	$10^{-5} x^2$	11.9×10^{-3}	2.74
Grey matter	0.4	$6 \times 10^{-5} x^2$	23×10^{-3}	53.61	$-3 \times 10^{-6} x^2$	-0.9×10^{-3}	0.82
	5	$6 \times 10^{-5} x^2$	20×10^{-3}	45.56	$4 \times 10^{-6} x^2$	0.6×10^{-3}	3.57
	10	$4 \times 10^{-5} x^2$	17×10^{-3}	40.10	$2 \times 10^{-6} x^2$	3.2×10^{-3}	7.93
ID	0.4	$0.8x^2 \times 10^{-3}$	0.25	62.15	$4 \times 10^{-5} x^2$	12.1×10^{-3}	1.39
	5	$0.7x^2 \times 10^{-3}$	0.23	51.92	$6 \times 10^{-5} x^2$	19.4×10^{-3}	5.10
	10	$0.6x^2 \times 10^{-3}$	0.22	44.28	$0.1x^2 \times 10^{-3}$	34.8×10^{-3}	11
IDC	0.4	$-0.2x^2 \times 10^{-3}$	26.5×10^{-3}	65.52	$9 \times 10^{-6} x^2$	2.4×10^{-3}	1.67
	5	$-0.3x^2 \times 10^{-3}$	52.3×10^{-3}	57.77	$-2 \times 10^{-5} x^2$	-3.2×10^{-3}	5.24
	10	$-0.3x^2 \times 10^{-3}$	36.9×10^{-3}	50.52	$0.1x^2 \times 10^{-3}$	-27.9×10^{-3}	11.70
Long bone	0.4	$0.7x^2 \times 10^{-3}$	0.24	29.73	$10^{-6} x^2$	4.3×10^{-3}	0.37
	5	$0.6x^2 \times 10^{-3}$	0.19	22.60	$6 \times 10^{-6} x^2$	18.6×10^{-3}	2.20
	10	$0.5x^2 \times 10^{-3}$	0.16	18.94	$0.1x^2 \times 10^{-3}$	39.7×10^{-3}	4.50
Skin	0.4	$-9 \times 10^{-6} x^2$	35.8×10^{-3}	47.54	$-2 \times 10^{-6} x^2$	0.2×10^{-3}	0.64
	5	$5 \times 10^{-6} x^2$	34.3×10^{-3}	38.82	$-2 \times 10^{-5} x^2$	-3.8×10^{-3}	3.34
	10	$5 \times 10^{-5} x^2$	47×10^{-3}	33.41	$-5 \times 10^{-5} x^2$	-10.9×10^{-3}	7.27
Skull	0.4	$0.4x^2 \times 10^{-3}$	0.19	44.61	$2 \times 10^{-6} x^2$	2.3×10^{-3}	0.62
	5	$0.3x^2 \times 10^{-3}$	0.17	35.96	$3 \times 10^{-5} x^2$	14.9×10^{-3}	3.34
	10	$0.3x^2 \times 10^{-3}$	0.14	30.40	$8 \times 10^{-5} x^2$	36.3×10^{-3}	7.25

(Continued)

Table 4.3 (Continued)

Tissue	Frequency (GHz)	Permittivity			Conductivity		
		A	B	C	A	B	C
Spinal cord	0.4	$0.3x^2 \times 10^{-3}$	0.13	39.05	$3 \times 10^{-6} x^2$	1.5×10^{-3}	0.50
	5	$0.3x^2 \times 10^{-3}$	0.11	32.69	$2 \times 10^{-6} x^2$	8×10^{-3}	2.50
	10	$0.3x^2 \times 10^{-3}$	99.2×10^{-3}	28.73	$5 \times 10^{-6} x^2$	19.5×10^{-3}	5.51
Tongue	0.4	$9 \times 10^{-5} x^2$	18.7×10^{-3}	55.84	$3 \times 10^{-6} x^2$	0.7×10^{-3}	0.89
	5	$7 \times 10^{-5} x^2$	12.1×10^{-3}	46.36	$10^{-6} x^2$	2.9×10^{-3}	4.42
	10	$4 \times 10^{-5} x^2$	6.8×10^{-3}	38.91	$3 \times 10^{-6} x^2$	5.8×10^{-3}	9.98
White matter	0.4	$0.5x^2 \times 10^{-3}$	0.17	42.95	$7 \times 10^{-6} x^2$	2.4×10^{-3}	0.57
	5	$0.5x^2 \times 10^{-3}$	0.16	36.03	$4 \times 10^{-5} x^2$	11.9×10^{-3}	2.86
	10	$0.4x^2 \times 10^{-3}$	0.14	31.47	$10^{-4} x^2$	30.8×10^{-3}	6.36

Table 4.4 Comparison between measured animal [2] permittivity and conductivity with Gabriel et al. data [3, 21] at **2450 MHz**, Where animal weight (10, 50, and 250 kg) corresponds to the properties of animal at specified weight.

Tissue	IFAC and ITIS [3, 21]		Peyman animal data [2]		
	ε'	σ (Sm^{-1})	ε'	σ (Sm^{-1})	Weight (kg)
Bone marrow 30%	5.30	0.10	5.61	0.11	250
Cornea	51.61	2.30	52.91	2.04	10
Dura	42.03	1.67	41.93	2.17	250
Fat	5.28	0.11	5.61	0.12	250
Grey matter	48.91	1.81	48.39	1.77	50
Intervertebral disc	39.70	1.66	42.84	2.06	250
Long bone (cortical)	11.38	0.39	13.03	0.46	250
Skin (dry)	38.00	1.46	41.26	1.60	50
Skull (cancellous)	18.5	0.81	17.2	0.64	250
Spinal cord	30.15	1.09	32.35	0.80	50
Tongue	52.63	1.80	52.37	2.05	250
White matter	36.17	1.22	37.59	1.31	10

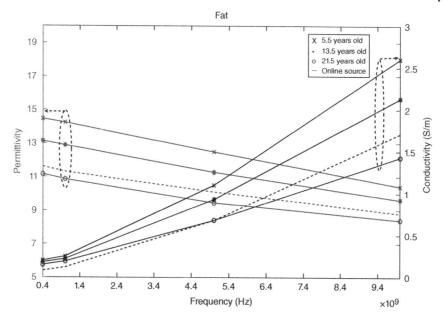

Figure 4.6 Comparison between age-dependent properties of fat tissue with online sources [3, 21], online sources dielectric properties well match with 250 kg pig when comparing with [2] as shown in Table 4.4 and also confirms by this method.

(0.4–10 GHz). In Figure 4.6, permittivity is presented along left Y-axis and conductivity along right Y-axis. The comparison is made between the data from proposed method (dispersive age-dependent method) and available online sources [3, 21]. The dielectric properties available on online sources were calculated from animals and human [22], and the factor of age dependences in measuring dielectric properties wasn't considered.

In Figure 4.6, a 21.5-year-old child dielectric properties (permittivity and conductivity) of fat tissue are very close to the online sources data that is in line with what Table 4.4 is predicting about the equivalence of fat tissue properties with 250 kg pig. In other words, fat tissue properties in online sources belong to adults.

This novel age-dependent method also indicates that there is difference in dielectric properties among the different ages (see Figure 4.6). Fat tissue properties indicate that at 1 GHz, for example, the permittivity value of a 5.5-year-old child is approx. 24% higher than 21.5-year-old human. Similarly, the conductivity is also approx. 28% higher as compared to 21.5-year-old human. Similar comparison has been made among different ages with different tissues types and also compared with literature data (online sources), which are presented in Figures 4.A.1–4.A.13.

It has been noticed that age-dependent tongue tissue properties are not in line with Table 4.4 predictions. According to this study, a 5.5-year-old (equivalent to 12.5 kg pig) child data is closest to the available online source data, whereas the previous table indicates that 250 kg pig is closest. The reason of such differences is unknown.

The percentage error has been calculated between single frequency and dispersive age-dependent methods in order to find the accuracy among the two methods. The error associated with permittivity and conductivity is well under one and ten percent, respectively, as mentioned before. It's worth mentioning that a very good agreement is achieved between two methods.

Furthermore, the Cole–Cole model has been applied in order to generate relaxation parameters which are AFD. The details about the implementation of the Cole–Cole model are explained in next section.

4.3.3.3 Implementation of the Cole–Cole Model on Age-Dependent Properties

According to the general relaxation theory, polar molecules in materials rearrange under the influence of applied E-field and contributing to the polarization. This phenomenon is known as dielectric relaxation [23]. It is well established that dielectric relaxation process is always frequency dependent. In accordance with frequency dependence, dielectric properties of a material are characterized by relaxation time constants.

The dielectric spectrum of biological tissues is composed of number of relaxation parameters. These parameters can be employed in a well-known mathematically model known as the Cole–Cole model or Cole–Cole expression [24] as given in Eq. (4.4).

$$\hat{\varepsilon}(\omega) = \varepsilon_\infty + \frac{\varepsilon_s - \varepsilon_\infty}{1 + (j\omega\tau)^{(1-\alpha)}} + \frac{\sigma_i}{j\omega\varepsilon_0} \qquad (4.4)$$

where $\hat{\varepsilon}$ is the complex relative permittivity, τ_{relax} is the relaxation time, ε_s is static dielectric constant, σ_i is the ionic conductivity, $0 \leq \alpha \leq 1$ is an adjustable parameter and indicating relaxation time distribution, and the Cole–Cole parameters have their usual significance.

Modelling the dielectric properties of tissues using Eq. (4.4) enables the incorporation in numerical assessments of human exposure to EMFs. In literature, the measured data about an animal was produced with the help of the Cole–Cole expression [2]. However, relaxation parameters for AFD dielectric properties are also calculated using this method (Cole–Cole).

In Eq. (4.4) all parameters were fitted except for the ε_∞ values. The value chosen for ε_∞ was 3, which is based on assumption similar to that reported in reference [2]. The Cole–Cole model parameters of 14 tissue types for a 5–22-year-old human are represented in Table 4.5. It's worth mentioning that all the parameters values including τ_{relax}, α, ε_s, and σ_i are very close as compared to reference [2].

The parameters listed in the previous table are based on pure empirical Eq. (4.4). This study is not intended to identify the underlying mechanistic mechanism. Moreover, the data for each tissue type was fitted separately at each age. The parameters listed in the previous table helps to calculate dielectric properties at any frequency (within 0.4–10 GHz) and age (5–22 years) without solving huge number of equations.

The relaxation parameters of bone marrow (as an example) are plotted in Figure 4.7 and values are fitted in order to show the trend with respect to the age of human. The static permittivity and ionic conductivity have been found main contributors in resultant dielectric properties while fitting the values in the Cole–Cole expression.

4.3.3.4 Accuracy Among the Age-dependent Methods

Dielectric properties of biological tissues play a vital role in SAR assessment. These new age-dependent techniques allow to increase the accuracy of SAR assessment when humans of various age groups exposed to EMFs. The percentage error has been calculated based on values generated by single frequency and dispersive age-dependent methods. The single frequency method produces dielectric values at specific frequency point, whereas the dispersive method generates over the whole frequency band (0.4–10 GHz). The values resultant from both methods were compared at specific frequency and age. The error associated with dispersive age-dependent method is well under 1 and 10% for both permittivity and conductivity, respectively. The single frequency age-dependent method is more accurate than the dispersive age-dependent method and also from the Cole–Cole model.

There is no consistent variation in dielectric properties w.r.t., age, and frequency. The variation in dielectric properties among the different ages for example between 5.5- and 7.5-year-old child is within 2–6% and similar variation noticed when compare different age groups. The highest variation observed in tissue bone marrow 30% was approx. 73% in permittivity and 85% in conductivity when 5.5-year-old child was compared with 21.5-year-old human, respectively. Some tissues dielectric properties increase with frequency and some shows the opposite trend. The uncertainty in tissue dielectric properties trend emphasizes to consider the AFD properties in order to improve the accuracy in SAR assessment.

Table 4.5 Cole–Cole parameters for the dielectric properties of human tissues as function of age.

Tissue/age	ε_s	$\tau_{relax\ (ps)}$	$\sigma_i\ (Sm^{-1})$	α
Bone marrow 30%				
5.5 years	35.66	8.42	0.53	0.28
13.5 years	15.06	10.72	0.14	0.24
21.5 years	9.41	11.50	0.07	0.15
Bone marrow 50%				
5.5 years	38.17	8.34	0.54	0.27
13.5 years	18.72	9.80	0.18	0.25
21.5 years	12.64	10.30	0.1	0.15
Cornea				
5.5 years	57.43	7.86	0.89	0.22
13.5 years	54.62	7.91	0.84	0.24
21.5 years	51.64	8.18	0.77	0.20
Dura				
5.5 years	57.43	7.86	0.89	0.22
13.5 years	54.62	7.91	0.84	0.24
21.5 years	51.64	8.18	0.77	0.20
Fat				
5.5 years	14.59	10.50	0.19	0.15
13.5 years	13.30	10.00	0.17	0.19
21.5 years	11.24	10.20	0.14	0.14
Grey matter				
5.5 years	55.12	6.23	0.77	0.34
13.5 years	54.16	6.20	0.8	0.34
21.5 years	52.63	6.83	0.86	0.32
Intervertebral disc				
5.5 years	60.78	7.81	1.17	0.28
13.5 years	52.99	9.18	0.71	0.26
21.5 years	46.28	10.80	0.0001	0.27

Table 4.5 (Continued)

Tissue/age	ε_s	$\tau_{\text{relax (ps)}}$	σ_i (Sm^{-1})	α
Intervertebral disc center				
5.5 years	66.66	6.20	1.59	0.21
13.5 years	66.92	6.62	1.51	0.16
21.5 years	66.82	6.50	1.36	0.20
Long bone				
5.5 years	28.79	14.00	0.27	0.38
13.5 years	20.97	18.50	0.11	0.38
21.5 years	12.91	18.50	0.08	0.36
Skin				
5.5 years	48.94	7.80	0.58	0.34
13.5 years	47.20	8.84	0.57	0.30
21.5 years	44.75	9.50	0.57	0.26
Skull				
5.5 years	43.94	8.84	0.53	0.33
13.5 years	37.37	9.37	0.45	0.35
21.5 years	29.39	10.50	0.33	0.33
Spinal cord				
5.5 years	39.02	6.21	0.44	0.37
13.5 years	34.9	6.54	0.38	0.38
21.5 years	29.73	7.00	0.31	0.33
Tongue				
5.5 years	56.94	8.35	0.82	0.26
13.5 years	56.17	8.22	0.79	0.26
21.5 years	54.77	8.71	0.76	0.22
White matter				
5.5 years	42.27	6.00	0.49	0.35
13.5 years	36.99	7.00	0.4	0.38
21.5 years	31.22	7.50	0.27	0.36

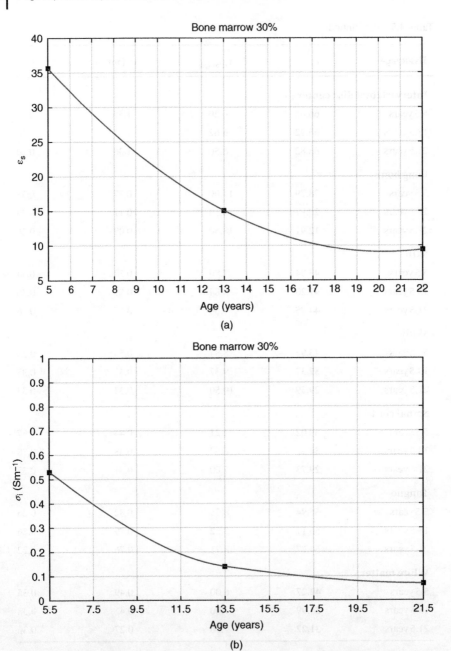

Figure 4.7 Age-dependent fitted values for relaxation parameters of the Cole–Cole model of bone marrow 30% tissue; (a) Static permittivity; (b) Ionic conductivity.

4.4 Numerical Validation

This section focuses on the validity and the accuracy of the methods, and the numerical tools used to compute the EM propagation and the absorptions characteristics within the body. Experimental validation of the FDTD algorithm has been conducted by a number of researchers [26–35]. In previously mentioned references, authors have computed the whole-body average SAR distribution in a variety of models including children and adults after being exposed to wide range of frequencies from a plane wave source.

Different scenarios of plane wave exposure have been considered in [26, 27, 31, 33]. Plane wave exposure scenarios include model standing in free space (suspended) or on ground plane. In reference [33], analysis has been made on SAR distribution inside human models by changing the posture of the model when exposed to EM waves. In reference [27], SAR assessment has been carried out on a sitting voxel model when exposed to variable frequencies using Wi-Fi devices. Plane wave exposures of voxel model and irradiation from antennas in the near field were studied for variety of exposure conditions. Similarly, another study [34] estimates the SAR exposure to EMFs from the antennas for wireless devices (Wi-Fi and Bluetooth) using FDTD methods. These studies have demonstrated a good agreement between SAR distributions identified numerically and experimentally.

In this study, further validations have been conducted by comparing the results obtained using FIT algorithm against analytical and other commercial EM computation packages using some simple benchmarks [26–30, 35]. A description of the main validation procedure is presented in this section.

4.4.1 Comparison with an Analytical Benchmark

A theoretical calculation model has been used as a benchmark for validating the FIT software. This theoretical model is used for expressing the EMFs and SAR induced inside a lossy dielectric sphere by an incident plane-wave. The analytical solutions was based on exposing single layered, muscle-equivalent model spheres, with radii of 3 and 15 cm, to the plane wave of 1 GHz radiation [36]. Similar dielectric properties of muscle tissue were used in analytical and computational models [37]. The theoretically predicted and computed SAR patterns are computed based on the patterns along the x, y, and z directions.

Plots of the normalized analytically assessed SAR patterns along the major axes of a muscle-equivalent dielectric sphere (radius = 3 cm) are shown in Figure 4.8. The black solid lines shows the analytically solved SAR pattern along the three major axes X, Y, and Z [36]. Calculations are based on the plane waves of 1 GHz, with energy flux density of $1 \, mWcm^{-2}$ and the WBSAR is $0.235 \, Wkg^{-1}$. For the

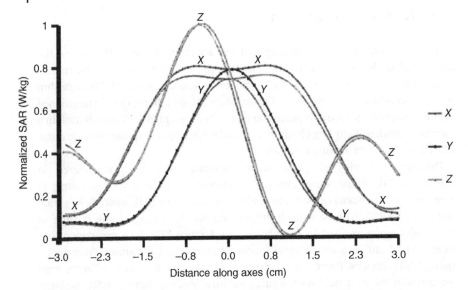

Figure 4.8 Plots of the normalized FIT-computed SAR patterns along the major axes of a muscle-equivalent dielectric sphere (Radius = 3 cm). Calculations are based on a plane wave source of **1 GHz**, with energy flux density of 1 mWcm^{-2}. WBSAR = 0.241 Wkg^{-1}.

convenience of presentation, the distributed SAR values are normalized to the maximum SAR given for each figure. It was found that strong standing wave patterns (hot spots) exist in the sphere.

Similarly, the computed SAR patterns inside the 3cm sphere using the CST MWS are also presented with colored lines in Figure 4.8. The SAR values have also been normalized to the maximum value. The dense mesh has used for this study where total mesh cells was 1.7 million with 15 lines per wavelength (inside the dielectric), which satisfies the step size and stability conditions for this frequency. The sphere models have been separated from the PML layers by 10 cells representing free space in order to ensure better absorbing boundary condition (ABC) performance. One of the disadvantages of fixing the number of surrounding free-space cells is the variability in separation distance between the tested object and the ABC, specifically when varying the size of the cell.

Based on the figure, an excellent correlation has been observed among the overall SAR pattern, which have been plotted from the analytical predictions and FIT (CST MWS) computations. The SAR values decreasing along the X, Y, and Z coordinates are noticed. However, these two graphs/results are in good comparison in order to validate this method. An excellent correlation has been observed among the overall estimated SAR, where error is found less than 3%.

4.5   Chapter Summary

This chapter summarizes modelling of three different voxel models, which were used to evaluate SAR due to exposure from wireless devices. Heterogeneous voxel models have been modelled using in-house modelling software package to handle voxels. Age- and frequency-dependent dielectric properties of 14 tissues have been calculated based on data fitting. The curve fitting techniques has been used and applied in three different methods on the available measured data of animals to predict the best fit dielectric properties for humans of various age groups. The Cole–Cole relaxation parameters have been also generated for all tissue type and very good agreement achieved with relaxation parameters of measured data.

Novel age-dependent methods are only applicable to extrapolate data for 5.5- to 21.5-year-old humans within frequency range of 0.4–20 GHz. The age and frequency limitations are solely due to the lack of availability of animal growth data and measured data at higher frequencies correspondingly. The newly calculated data was then used to assess the induced SAR in a female child model [8] when exposed to a plane waves and smart meter devices in Chapter 4. Moreover, the numerical tool (which is based on finite integration technique (FIT) method) was also validated by comparing the FIT calculations over a homogenous muscle-equivalent sphere with analytical Mie series solution for homogenous sphere.

Appendix 4.A

Table 4.A.1   Dielectric properties of 40 different tissue types assigned to NAOMI at **868 MHz** and **2450 MHz**, where the dielectric data is chosen from various sources to improve accuracy of assessment.

Tissue types	Frequency 868 MHz		Frequency 2450 MHz		Mass density (kg/m³)	Source
	Relative permittivity	Conductivity (S/m)	Relative permittivity	Conductivity (S/m)		
Adrenal gland	50.5	1.00	48.2	1.77	1028	Peyman and Gabriel [7]
Air (except lungs)	1.0	0.0	1.0	0.0	1	IFAC
Background	1.0	0.0	1.0	0.0	0	IFAC
Bile	70.3	1.83	68.4	2.79	928	IFAC
Blood	61.5	1.52	58.3	2.54	1050	IFAC

(Continued)

Table 4.A.1 (Continued)

Tissue types	Frequency 868 MHz		Frequency 2450 MHz		Mass density (kg/m^3)	Source
	Relative permittivity	Conductivity (S/m)	Relative permittivity	Conductivity (S/m)		
Bone cancellous	20.9	0.33	18.6	0.80	1178	IFAC
Bone cortical	12.5	0.14	11.4	0.39	1908	IFAC
Brain	46.0	0.76	42.6	1.49	1043	IFAC
Breast fat	25.9	0.49	24.5	0.93	911	IFAC and Peyman and Gabriel [2]
Cartilage	42.8	0.77	38.8	1.75	1100	IFAC
Cerebrospinal fluid	68.7	2.40	66.3	3.45	1007	IFAC
Duodenum	65.2	1.17	62.2	2.20	1030	IFAC
Eye lens	46.6	0.78	44.6	1.50	1076	IFAC
Eye sclera	55.4	1.15	52.6	2.03	1032	IFAC
Fallopian ovaries	50.7	1.27	44.7	2.26	1048	IFAC
Fat	5.5	0.05	5.3	0.10	911	IFAC
Gall bladder	59.2	1.25	57.6	2.05	1071	IFAC
Heart muscle	60.1	1.21	54.8	2.25	1081	IFAC
Kidney	59.0	1.37	52.8	2.42	1066	IFAC
Liver	47.0	0.84	43.1	1.68	1079	IFAC
Lower large intestine	59.7	2.15	54.4	3.17	1000	IFAC
Lunch	65.2	1.17	62.2	2.20	1088	IFAC
Lung	22.1	0.45	20.5	0.80	394	IFAC
Muscle	55.1	0.93	52.7	1.73	1090	IFAC
Oesophagus	65.2	1.17	62.2	2.20	1040	IFAC
Pancreas	59.8	1.03	57.2	1.96	1087	IFAC
Skin	41.6	0.86	38.0	1.46	1109	IFAC
Small intestine	59.7	2.15	54.4	3.17	1030	IFAC
Spinal cord	32.6	0.56	30.2	1.09	1075	IFAC
Spleen	57.4	1.26	52.5	2.05	1089	IFAC

(Continued)

Table 4.A.1 (Continued)

Tissue types	Frequency 868 MHz		Frequency 2450 MHz		Mass density (kg/m³)	Source
	Relative permittivity	Conductivity (S/m)	Relative permittivity	Conductivity (S/m)		
Stomach	65.2	1.17	62.2	2.20	1088	IFAC
Tendon	45.9	0.71	43.1	1.68	1142	IFAC
Thymus	59.8	1.03	57.2	1.96	1023	IFAC
Thyroid gland	59.8	1.03	57.2	1.96	1050	IFAC
Upper large intestine	59.7	2.15	54.4	3.17	1000	IFAC
Urinary bladder	59.2	1.25	18.0	0.68	1035	IFAC
Urine	68.7	2.01	67.8	2.83	1035	Peyman and Gabriel [7]
Uterus	61.2	1.26	57.8	2.24	1105	IFAC
Vagina	50.7	1.27	44.7	2.24	1048	IFAC
Vitreous humour	68.9	1.63	68.2	2.47	1005	IFAC

Table 4.A.2 Dielectric properties of 38 different tissue types assigned to NORMAN at **868 MHz** and **2450 MHz**, where the dielectric data is chosen from various sources to improve the accuracy of the assessment.

Tissue types	Frequency 868 MHz		Frequency 2450 MHz		Mass density (kg/m³)	Source
	Relative permittivity	Conductivity (S/m)	Relative permittivity	Conductivity (S/m)		
Adrenal gland	50.5	1.00	48.2	1.77	1028	Peyman and Gabriel [7]
Air (except lungs)	1.0	0.00	1.0	0.00	1	IFAC
Background	1.0	0.00	1.0	0.00	0	IFAC
Bile	70.3	1.83	68.4	2.79	928	IFAC
Blood	61.5	1.52	58.3	2.54	1050	IFAC
Bone cancellous	20.9	0.33	18.6	0.80	1178	IFAC

(Continued)

Table 4.A.2 (Continued)

Tissue types	Frequency 868 MHz		Frequency 2450 MHz		Mass density (kg/m³)	Source
	Relative permittivity	Conductivity (S/m)	Relative permittivity	Conductivity (S/m)		
Bone cortical	12.5	0.14	11.4	0.39	1908	IFAC
Brain	45.9	0.76	42.6	1.49	1046	IFAC
Breast fat	25.9	0.49	24.5	0.93	911	IFAC and Peyman and Gabriel [2]
Cerebrospinal fluid	68.7	2.40	66.3	3.45	1007	IFAC
Duodenum	65.2	1.17	62.2	2.20	1030	IFAC
Eye lens	46.6	0.78	44.6	1.50	1076	IFAC
Eye sclera	55.4	1.15	52.6	2.03	1032	IFAC
Fat	5.5	0.05	5.3	0.10	911	IFAC
Gall bladder	59.2	1.25	57.6	2.05	1071	IFAC
Heart muscle	60.1	1.21	54.8	2.25	1081	IFAC
Kidney	59.0	1.37	52.8	2.42	1066	IFAC
Liver	47.0	0.80	43.1	1.70	1079	IFAC
Lower large intestine	59.7	2.15	54.4	3.17	1000	IFAC
Lunch	65.2	1.17	62.2	2.20	1088	IFAC
Lung	22.1	0.45	20.5	0.80	394	IFAC
Muscle	55.1	0.93	52.7	1.73	1090	IFAC
Oesophagus	65.2	1.17	62.2	2.20	1040	IFAC
Pancreas	59.8	1.03	57.2	1.96	1087	IFAC
Prostate	60.7	1.20	57.6	2.16	1045	IFAC
Skin	41.6	0.86	38.0	1.46	1109	IFAC
Small intestine	59.7	2.15	54.4	3.17	1030	IFAC
Spinal cord	32.6	0.56	30.2	1.09	1075	IFAC
Spleen	57.4	1.26	52.5	2.05	1089	IFAC
Stomach	65.2	1.17	62.2	2.20	1088	IFAC
Tendon	45.9	0.71	43.1	1.68	1142	IFAC
Testis	60.7	1.20	57.6	2.16	1082	IFAC

(Continued)

Table 4.A.2 (Continued)

Tissue types	Frequency 868 MHz		Frequency 2450 MHz		Mass density (kg/m³)	Source
	Relative permittivity	Conductivity (S/m)	Relative permittivity	Conductivity (S/m)		
Thymus	59.8	1.03	57.2	1.96	1023	IFAC
Thyroid gland	59.8	1.03	57.2	1.96	1050	IFAC
Upper large intestine	59.7	2.15	54.4	3.17	1000	IFAC
Urinary bladder	59.2	1.25	18.0	0.68	1035	IFAC
Urine	68.7	2.01	67.8	2.83	1035	Peyman and Gabriel [2]
Vitreous humour	68.9	1.63	68.2	2.47	1005	IFAC

Table 4.A.3 Dielectric properties of 75 different tissue types of Eartha at **868 MHz** and **2450 MHz**, where the dielectric data is chosen from various sources for accurate assessment (IE21P: Interpolation Equations from 21.3 kg pig).

Tissue names	Frequency 868 MHz		Frequency 2450 MHz		Mass density (kg/m³)	Assumptions	Source
	Relative permittivity	Conductivity (S/m)	Relative permittivity	Conductivity (S/m)			
Adrenal gland	50.5	1.00	48.2	1.77	1028	Adrenals	IFAC
Air internal	1.0	1.00	1.0	0.00	1	Air (except lungs)	IFAC
Artery	61.5	1.52	58.3	2.54	1050	Blood	IFAC
Bladder	59.2	1.25	18.0	0.68	1035	Bladder	IFAC
Blood vessel	61.5	1.52	58.3	2.54	1050	Blood	IFAC
Bone	22.6	0.35	20.1	0.85	1908	Long bone	IE21P
Brain grey matter	44.6	0.86	41.8	1.49	1043	50% grey and white matter = brain	IE21P
Brain white matter	44.6	0.86	41.8	1.49	1043	50% grey and white matter = brain	IE21P

(Continued)

Table 4.A.3 (Continued)

Tissue names	Frequency 868 MHz		Frequency 2450 MHz		Mass density (kg/m³)	Assumptions	Source
	Relative permittivity	Conductivity (S/m)	Relative permittivity	Conductivity (S/m)			
Bronchi	42.1	0.76	39.7	1.45	1102	Bronchi	IT'IS
Bronchi lumen	1.0	0.00	1.0	0.00	1	Bronchi lumen	IT'IS
Cartilage	42.8	0.77	38.8	1.75	1100	Cartilage	IFAC
Cerebellum	49.7	1.25	44.8	2.10	1045	Cerebellum	IT'IS
CSF	68.7	2.40	66.3	3.45	1007	CSF	IFAC
Commissura anterior	39.0	0.58	36.2	1.21	1041	Commissura anterior	IT'IS
Commissura posterior	39.0	0.58	36.2	1.21	1041	Commissura Posterior	IT'IS
Connective tissue	45.9	0.71	43.1	1.68	1027	Connective tissue	IT'IS
Cornea	54.5	1.07	51.6	3.54	1051	Cornea	IE21P
Diaphragm	55.1	0.93	52.7	1.74	1090	Diaphragm	IT'IS
Ear cartilage	42.8	0.77	38.8	1.75	1100	Cartilage	IFAC
Ear skin	45.50	0.79	42.2	1.64	1109	Skin	IE21P
Eye lens	46.6	0.78	44.6	1.50	1076	Eye Lens	IT'IS
Eye sclera	55.4	1.15	52.6	2.03	1032	Eye Sclera	IT'IS
Eye vitreous humour	68.9	1.63	68.2	2.47	1005	Eye Vitreous Humour	IT'IS/IFAC
Fat	14.0	0.23	13.3	0.46	911	Fat	IE21P
Gallbladder	59.2	1.25	57.6	2.05	1071	Gallbladder	IFAC
Heart lumen	61.5	1.52	58.3	2.54	1050	Heart lumen	IT'IS
Heart muscle	60.1	1.21	54.8	2.25	1081	Heart muscle	IT'IS
Hippocampus	44.6	0.86	41.8	1.49	1043	Brain	IE21P
Hypophysis	59.8	1.03	57.2	1.97	1053	Hypophysis	IT'IS
Hypothalamus	59.8	1.03	57.2	1.97	1053	Hypothalamus	IT'IS
Intervertebral disc	55.1	1.28	51.4	2.29	1100	Intervertebral disc	IE21P
Kidney cortex	58.9	1.37	52.7	2.43	1049	Kidney cortex	IT'IS
Kidney medulla	58.0	1.37	52.7	2.43	1044	Kidney medulla	IT'IS

(Continued)

Table 4.A.3 (Continued)

Tissue names	Frequency 868 MHz		Frequency 2450 MHz		Mass density (kg/m³)	Assumptions	Source
	Relative permittivity	Conductivity (S/m)	Relative permittivity	Conductivity (S/m)			
Large intestine	58.1	1.07	53.9	2.04	1088	Large intestine	IT'IS
Large intestine lumen	55.1	0.93	52.7	1.74	1045	Large intestine lumen	IT'IS
Larynx	42.8	0.77	38.8	1.75	1100	Larynx	IT'IS
Liver	47.0	0.80	43.1	1.70	1079	Liver	IFAC
Lung	22.1	0.45	20.5	0.80	394	Lung	IFAC
Mandible	12.5	0.14	11.4	0.39	1908	Mandible	IT'IS
Marrow red	11.3	0.22	10.3	0.45	1029	Marrow red	IT'IS
Medulla oblongata	49.7	1.25	44.8	2.10	1046	Medulla oblongata	IT'IS
Meniscus	42.8	0.76	38.8	1.76	1100	Meniscus	IT'IS
Midbrain	49.7	1.25	44.8	2.10	1046	Midbrain	IT'IS
Mucosa	55.1	0.93	52.7	1.74	1102	Mucosa	IT'IS
Muscle	55.1	0.9	52.7	1.7	1090	muscle	IFAC
Nerve	32.6	0.56	30.1	1.09	1075	Nerve	IT'IS
Oesophagus	65.2	1.17	62.2	2.20	1040	Oesophagus	IFAC
Oesophagus lumen	1	0.0	1.0	0.0	1	Oesophagus lumen	IT'IS
Ovary	50.7	1.27	44.7	2.26	1100	Fallopian ovaries	IFAC
Pancreas	59.8	1.03	57.2	1.96	1087	Pancreas	IFAC
Patella	45.9	0.71	43.1	1.68	1142	Tendon	IFAC
Pharynx	1.0	0.0	1.0	0.0	1	Pharynx	IT'IS
Pineal body	59.8	1.03	57.2	1.97	1053	Pineal body	IT'IS
Pons	49.7	1.25	44.8	2.10	1046	Pons	IT'IS
SAT (Subcutaneous Fat)	11.3	0.10	10.8	0.26	911	Subcutaneous fat	IT'IS
Skin	45.50	0.79	42.2	1.64	1079	Skin	IE21P
Skull	39.3	0.71	36.1	1.43	1850 [36]	Skull	IE21P

(Continued)

Table 4.A.3 (Continued)

Tissue names	Frequency 868 MHz		Frequency 2450 MHz		Mass density (kg/m^3)	Assumptions	Source
	Relative permittivity	Conductivity (S/m)	Relative permittivity	Conductivity (S/m)			
Small intestine	59.7	2.15	54.4	3.17	1030	Small intestine	IFAC
Small intestine lumen	55.1	0.93	52.7	1.74	1045	Small intestine lumen	IT'IS
Spinal cord	66.30	1.76	63.1	2.71	1075	Spinal cord	IE21P
Spleen	57.4	1.26	52.5	2.05	1089	Spleen	IFAC
Stomach	65.2	1.17	62.2	2.20	1088	Stomach	IFAC
Stomach lumen	55.1	0.93	52.7	1.74	1045	Stomach	IT'IS
Teeth	12.5	0.1	11.4	0.39	2180	Tooth	IT'IS
Tendon ligament	45.9	0.71	43.1	1.68	1142	Tendon	IFAC
Thalamus	52.9	0.92	48.9	1.81	1045	Thalamus	IT'IS
Thymus	59.8	1.03	57.2	1.96	1023	Thymus	IFAC
Tongue	54.2	1.03	50.9	2.01	1090	Tongue	IT'IS
Trachea	42.1	0.76	39.7	1.45	1080	Trachea	IT'IS
Trachea lumen	1.0	0.0	1.0	0.0	1	Trachea lumen	IT'IS
Ureter\urethra	44.9	0.68	42.5	1.44	1102	Ureter\Urethra	IT'IS
Uterus	61.2	1.26	57.8	2.25	1105	Uterus	IT'IS
Vagina	59.7	2.15	54.4	3.17	1088	Vagina	IT'IS
Vein	61.5	1.52	58.3	2.54	1050	Blood	IFAC
Vertebrae	55.1	1.28	51.4	2.29	1100	Intervertebral disc	IE21P

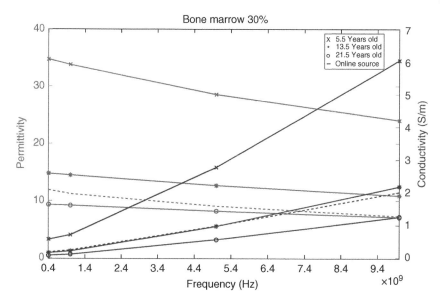

Figure 4.A.1 Comparison between age-dependent properties of bone marrow 30% tissue with online sources [3, 21], online sources dielectric properties well match with 250 kg pig when comparing with [2] as shown in Table 4.4 and also confirms by this method.

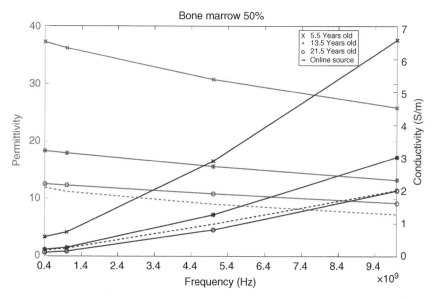

Figure 4.A.2 Comparison between age-dependent properties of bone marrow 50% tissue with online sources [3, 21], online sources dielectric properties well match with 250 kg pig when comparing with [2] and also confirms by this method.

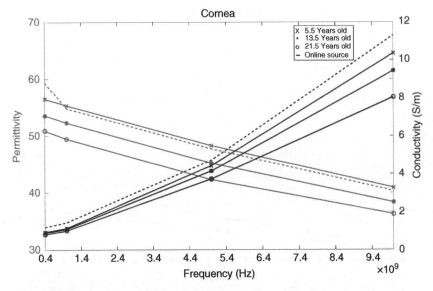

Figure 4.A.3 Comparison between age-dependent properties of cornea tissue with online sources [3, 21], online sources dielectric properties well match with 10 kg pig when comparing with [2] as shown in Table 4.4 and also confirms by this method.

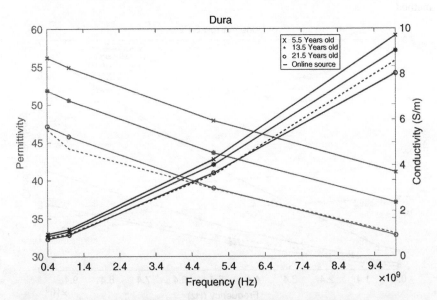

Figure 4.A.4 Comparison between age-dependent properties of dura tissue with online sources [3, 21], online sources dielectric properties well match with 250 kg pig when comparing with [2] as shown in Table 4.4 and also confirms by this method.

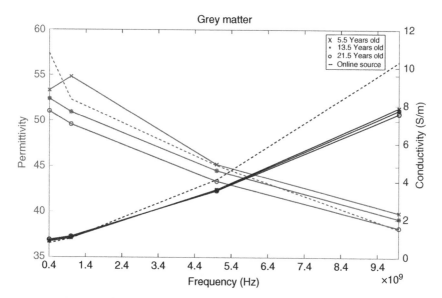

Figure 4.A.5 Comparison between age-dependent properties of gray matter tissue with online sources [3, 21], online sources dielectric properties well match with 50 kg pig when comparing with [2] as shown in Table 4.4 and also confirms by this method.

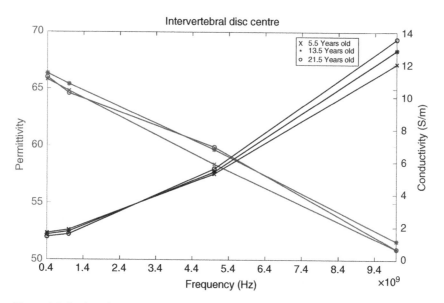

Figure 4.A.6 Age-dependent properties of intervertebral disc centre. Due to lack of availability of online source data comparison is not available of this tissue.

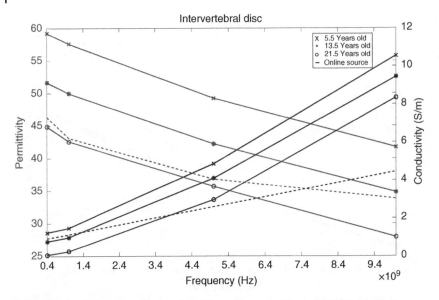

Figure 4.A.7 Comparison between age-dependent properties of intervertebral disc tissue with online sources [3, 21], online sources dielectric properties well match with 250 kg pig when comparing with [2] as shown in Table 4.4 and also confirms by this method.

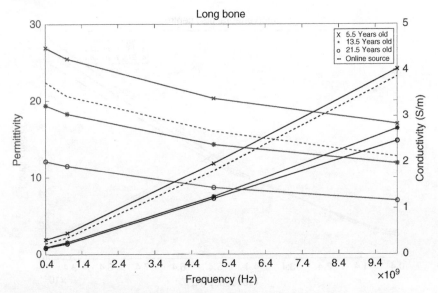

Figure 4.A.8 Comparison between age-dependent properties of long bone tissue with online sources [3, 21], online sources dielectric properties well match with 250 kg pig when comparing with [2] as shown in Table 4.4 and also confirms by this method.

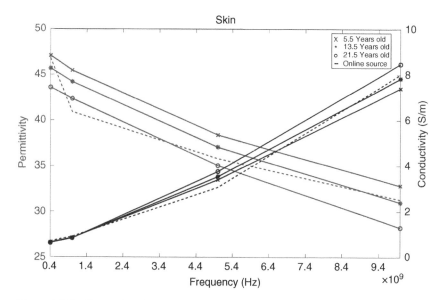

Figure 4.A.9 Comparison between age-dependent properties of skin tissue with online sources [3, 21], online sources dielectric properties well match with 50 kg pig when comparing with [2] as shown in Table 4.4 and also confirms by this method.

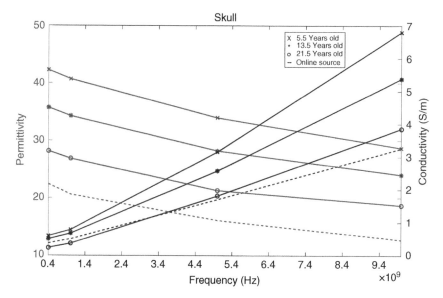

Figure 4.A.10 Comparison between age-dependent properties of skull tissue with online sources [3, 21], online sources dielectric properties well match with 250 kg pig when comparing with [2] as shown in Table 4.4 and also confirms by this method.

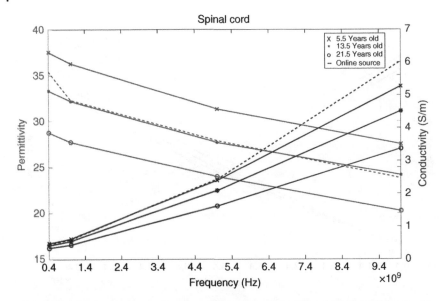

Figure 4.A.11 Comparison between age-dependent properties of spinal cord tissue with online sources [3, 21], online sources dielectric properties well match with 50 kg pig when comparing with [2] as shown in Table 4.4 and also confirms by this method.

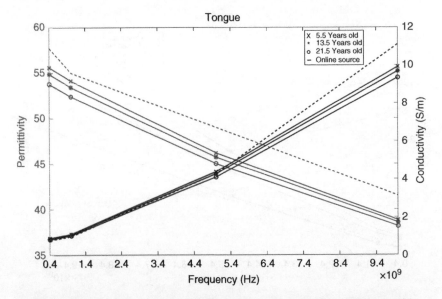

Figure 4.A.12 Comparison between age-dependent properties of tongue tissue with online sources [3, 21], online sources dielectric properties well match with 250 kg pig when comparing with [2] as shown in Table 4.4 and also confirms by this method.

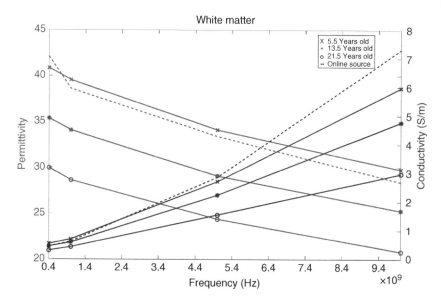

Figure 4.A.13 Comparison between age-dependent properties of white matter tissue with online sources [3, 21], online sources dielectric properties well match with 10 kg pig when comparing with [2] as shown in Table 4.4 and also confirms by this method.

References

1 CST Microwave Studio - 3D EM simulation software (2016). https://www.cst.com/Products/CSTMWS (accessed June 2016).

2 Peyman, A. and Gabriel, C. (2010). Cole–Cole parameters for the dielectric properties of porcine tissues as a function of age at microwave frequencies. *Physics in Medicine and Biology* 55: 413–419.

3 IFAC. Dielectric properties of body tissues. http://niremf.ifac.cnr.it/tissprop/htmlclie/htmlclie.htm (accessed February 2015).

4 Gabriel, C., Gabriely, S., and Corthout, E. (1996). The dielectric properties of biological tissues: I. Literature survey. *Physics in Medicine and Biology* 41: 2231–2249.

5 Gabriel, S., Lau, R.W., and Gabriel, C. (1996). The dielectric properties of biological tissues: II. Measurements in the frequency range 10 Hz to 20 GHz. *Physics in Medicine and Biology* 41: 2251–2269.

6 Gabriel, S., Lau, R.W., and Gabriel, C. (1996). The dielectric properties of biological tissues: III. Parametric models for the dielectric spectrum of tissues. *Physics in Medicine and Biology* 41: 2271–2293.

7 Peyman, A. and Gabriel, C. (2012). Dielectric properties of porcine glands, gonads and body fluids. *Physics in Medicine and Biology* 57: 339–344.

8 IT'IS, High-resolution human models for simulations: virtual population. http://www.itis.ethz.ch/itis-for-health/virtual-population/human-models/ (accessed February 2017).

9 Peyman, A., Rezazadeh, A.A., and Gabriel, C. (2001). Changes in the dielectric properties of rat tissue as a function of age at microwave frequencies. *Physics in Medicine and Biology* 46: 1617–1629.

10 The pig site (2014). http://www.thepigsite.com/stockstds/17/growth-rate (accessed August 2014).

11 Fukagawa, N.K., Bandini, L.G., Dietz, W.H., and Young, J.B. (1996). Effect of age on body water and resting metabolic rate. *Journal of Gerontology: Medical Sciences* 51: 71–73.

12 Bakker, J.F., Paulides, M.M., Christ, A. et al. (2010). Assessment of induced SAR in children exposed to electromagnetic plane waves between 10 MHz and 5.6 GHz. *Physics in Medicine and Biology* 55: 3115–3130.

13 Piuzzi, E., Bernardi, P., Cavagnaro, M. et al. (2011). Analysis of adult child exposure to uniform plane waves at mobile communication systems frequencies (900 MHz–3 GHz). *IEEE Transactions on Electromagnetic Compatibility* 53: 38–47.

14 Thurai, M., Goodridge, V.D., Sheppard, R.J., and Grant, E.H. (1984). Variation with age of the dielectric properties of mouse brain cerebrum. *Physics in Medicine and Biology* 29: 1133–1136.

15 Thurai, M., Steel, M.C., Sheppard, R.J., and Grant, E.H. (1985). Dielectric properties of developing rabbit brain at 37 degrees C. *Bioelectromagnetics* 6: 235–242.

16 Quinn, R. (2005). Comparing rat's to human's age: How old is my rat in people years? *Nutrition* 21: 775–777.

17 Peyman, A., Holden, S.J., Watts, S. et al. (2007). Dielectric properties of porcine cerebrospinal tissues at microwave frequencies: in vivo, in vitro and systematic variation with age. *Physics in Medicine and Biology* 52: 2229–2245.

18 Peyman, A., Gabriel, C., Grant, E.H. et al. (2009). Variation of the dielectric properties of tissues with age: the effect on the values of SAR in children when exposed to walkie–talkie devices. *Physics in Medicine and Biology* 54: 227–241.

19 Swindle, M.M., Makin, A., Herron, A.J. et al. (2012). Swine as models in biomedical research and toxicology testing. *Veterinary Pathology* 49: 344–356.

20 Valentin, J. (2002). Basic anatomical and physiological data for use in radiological protection: reference values: ICRP Publication 89. *Annals of the ICRP* 32: 1–277.

21 IT'IS foundation tissue properties. http://www.itis.ethz.ch/virtual-population/tissue-properties/database/database-summary/ (accessed February 2015).

22 Gabriel, C. (1996). Compilation of the dielectric properties of body tissues at RF and microwave frequencies, Final Technical Report. SPONSORING/MONITORING AGENCY REPORT NUMBER AL/OE-TR-1996-0004.

23 Kuang, W. and Nelson, S.O. (1998). Low-frequency dieletric properties of biological tissues: a review with some insights. *Transactions of ASAE* 41: 173–184.

24 Peyman, A., Holden, S., and Gabriel, C. (2005). Dielectric properties of tissues at microwave frequencies, RUM3 MTHR Final Technical Report.

25 Hirata, A., Yanase, K., Laakso, I. et al. (2012). Estimation of the whole-body averaged SAR of grounded human models for plane wave exposure at respective resonance frequencies. *Physics in Medicine and Biology* 57: 8427–8442.

26 Dimbylow, P.J. (2002). Fine resolution calculations of SAR in the human body for frequencies up to 3 GHz. *Physics in Medicine and Biology* 47: 2835–2846.

27 Findlay, R.P. and Dimbylow, P.J. (2010). SAR in a child voxel phantom from exposure to wireless computer networks (Wi-Fi). *Physics in Medicine and Biology* 55: 405–411.

28 Bakker, J.F., Paulides, M.M., Christ, A. et al. (2010). Assessment of induced SAR in children exposed to electromagnetic plane waves between 10 MHz and 5.6 GHz. *Physics in Medicine and Biology* 55: 3115–3130.

29 Christ, A., Kainz, W., Hahn, E.G. et al. (2010). The Virtual Family—development of surface-based anatomical models of two adults and two children for dosimetric simulations. *Physics in Medicine and Biology* 55: 23–38.

30 Hirata, A., Fujiwara, O., Nagaoka, T., and Watanabe, S. (2010). Estimation of whole-body average SAR in human models due to plane-wave exposure at resonance frequency. *IEEE Transactions on Electromagnetic Compatibility* 52: 41–48.

31 Dimbylow, P.J. (1997). FDTD calculations of the whole-body averaged SAR in an anatomically realistic voxel model of the human body from 1 MHz to 1 GHz. *Physics in Medicine and Biology* 42: 479–490.

32 Dimbylow, P.J. and Bolch, W. (2007). Whole-body-averaged SAR from 50 MHz to 4 GHz in the University of Florida child voxel phantoms. *Physics in Medicine and Biology* 52: 6639–6649.

33 Findlay, R.P. and Dimbylow, P.J. (2005). Effects of posture on FDTD calculations of specific absorption rate in a voxel model of human body. *Physics in Medicine and Biology* 50: 3825–3835.

34 Martínez-Búrdalo, M., Martín, A., Sanchis, A., and Villar, R. (2009). FDTD assessment of human exposure to electromagnetic fields from WiFi and bluetooth devices in some operating situations. *BEMS* 30: 142–151.

35 Hurt, W.D., Ziriax, J.M., and Mason, P.A. (2000). Variability in EMF permittivity values: implications for SAR calculations. *IEEE Transactions on Biomedical Engineering* 47: 396–401.

36 Ho, H.S. and Guy, A.W. (1974). Development of dosimetry For RF and microwave radiation-II calculations of absorbed dose distribution in two sizes of muscle-equivalent spheres. *Health Physics Journal* 29: 317–324.

37 Schwan, H.P. and Piersol, G.M. (1954). The absorption of electromagnetic energy in body tissues. *American Journal of Physical Medicine* 33: 371–404.

5

Antenna Design Considerations for Low SAR Mobile Terminals

Muhammad Ali Jamshed[1], Tim W.C. Brown[2], and Fabien Héliot[2]

[1]*James Watt School of Engineering, University of Glasgow, Glasgow, UK*
[2]*Institute of Communication Systems (ICS), Home of 5G and 6G Innovation Centre, University of Surrey, Guildford, UK*

5.1 Introduction

For more than three decades, mobile user equipment (UE) devices have advanced in terms of design and compactness. As their use has increased, rules have been placed in order to restrict the level of EM field (EMF) coupled to the human head, as measured by the specific absorption rate (SAR) [1]. In the beginning, mobile devices were either "candy bar" or "clam shell" in shape, and the emphasis was on how the external whip antenna could be positioned, or an interior planar antenna could be constructed to decrease SAR by utilizing the ground plane [2]. Over the last decade, the growing use of the "smartphone" has necessitated a rethinking of how the SAR may be significantly decreased for a significantly broader, yet thinner, cellular device. All recent smartphones are built around multiple-input multiple-output (MIMO) transceiver technology, which is needed by the 4th and 5th generation of cellular systems [3] as well as WiFi standards [4]. In general, the multiple antennas in a smartphone device are optimized first and foremost for efficiency and MIMO performance, whereas SAR is merely viewed as a regulatory constraint and, as such, is frequently an afterthought of the MIMO antenna design process [5, 6]. Early works related to reducing SAR in UEs have investigated ferrites [7, 8], defected ground structures [9], parasitic elements [10], or metamaterials [11, 12] as ways to decouple the UE antennas from the head, but the already very limited space in a smartphone makes these techniques almost impossible to implement.

The relative phase angle between MIMO antennas has also been discovered as a factor influencing SAR [13–15]. More specifically, it was demonstrated in [13] and [14] that using a suitable relative phase angle between the antenna elements

Low Electromagnetic Field Exposure Wireless Devices: Fundamentals and Recent Advances, First Edition.
Edited by Masood Ur Rehman and Muhammad Ali Jamshed.

could significantly reduce the SAR, and [14] proposed a signal processing scheme to implement this finding. In the [16], an SAR study was performed on a variety of UEs with MIMO antennas and demonstrated that the maximum SAR owing to relative phase angle between antenna components can be anticipated, which may be used to expedite conformance tests. This research also discovered that the SAR changes substantially with phase when considering closely spaced antennas, and that the power distribution between the antenna elements can alter the SAR, as previously discovered in [17]. Another research has found that the balance of power distribution between two components has an additional effect [18]. Though relative phase and power distribution between antenna elements are obvious criteria for modifying SAR, the fundamental physical basis of this phenomenon has not been adequately described and understood. This will allow for the development of context-aware multiple antennas for mobile terminals that can adapt to low SAR when in talk mode.

This chapter analyzes how to arrange two antenna components on a mobile terminal to generate the lowest achievable SAR in talk position while retaining efficiency. This is accomplished by power distribution between components, superposition of fields penetrating into the head, but also change in current distribution owing to coupling between elements, such that relative phase between elements manipulates and decreases the SAR to the human head. It does a thorough analysis of the antenna element needs, needed element coupling, and the impact of changing the inclination angle and earpiece location relative to the ear. Through antenna prototyping, it is confirmed that the coupling levels are also sufficiently low that MIMO capacity is not significantly impaired when employed away from the head. However, unlike traditional MIMO antennas, capacity is a secondary goal in this work, with SAR coming first. The original contributions to knowledge give a systematic examination of how two UE antennas and the ground plane may be used to minimize SAR to the human head while maintaining efficiency by optimizing their relative phase and coupling. The low SAR is maintained independent of the dielectric composition of the head, the inclination angle of the UE, or the location of the earpiece relative to the ear hole. Hand grip affects enhance SAR to the hand rather than the head.

The rest of the chapter is structured as follows: Section 5.2 delves into the theory underpinning antenna inter element coupling and changes in SAR. As a result, the comprehensive simulations presented in Section 5.3 are required to analyze the change in SAR and efficiency owing to relative phase from two coupled planar inverted-F (PIFA) components. Section 5.4 investigates the SAR using power splitting, field penetration, and current distributions. Section 5.5 investigates the concept's robustness in terms of inclination angle, position, and hand grip. Section 5.6 presents experimental data to validate the effect on MIMO capacity while not in talk position. Section 5.7 represents the chapter's conclusion.

5.2   SAR Reduction and Dual Coupling of Antenna

The term "mutual coupling" refers to the transfer of electromagnetic energy between any two antenna elements positioned close to one another [19]. This transfer is measured by the scattering parameter S_{21} (or S_{12}), which has the following relationships to the admittance parameters [20]:

$$S_{12} = \frac{-2Y_{12}Z_0^2}{(1 + Y_{11}Z_0)(1 + Y_{22}Z_0) - Y_{12}Y_{21}Z_0^2} \tag{5.1}$$

where Z_0 is the characteristic impedance, Y_{22} and Y_{11} are self-admittances at ports 2 and 1, respectively, and $Y_{21} = Y_{12}$ when reciprocity is assumed. By selecting Y_{12}, it is possible to link it to the electric field in the area of the antennas at frequency f using the following integral equation [20]:

$$Y_{12}(f) = \frac{1}{V_1 V_2} \iiint_{V'} (\mathbf{E}_1(r_2; f).\mathbf{J}_2(r_2; f)) \, dV', \tag{5.2}$$

where V_2 and V_1 represent the excited voltages of each antenna, and $\mathbf{E}_1$ represents the intensity of the radiated electric field of antenna 1 to the area r_2 of antenna 2 inside a three-dimensional volume V'. Based on the infinitesimal dipole model, this produces a current distribution, $\mathbf{J}_2$, at antenna 2. The electric field from antenna 2 creates a current dispersion at antenna 1. When both antennas radiate at the same time, the coupled current density to the opposite element and ground plane, as well as the driven current of each antenna, superimposes on each other, significantly altering the current distribution of the entire UE, which is dependent on the relative phase between elements.

According to the findings in [21], the current distribution of an antenna will have a direct impact on the resulting tangential magnetic field H_{tp} and penetrating electric field E_p into the surface of the head, where brain tissue has a high dielectric constant, ϵ_r:

$$E_p = \left(\frac{\mu_0}{\epsilon_0 \epsilon_r} \right) H_{tp} \tag{5.3}$$

where ϵ_0 and μ_0 are the permittivity and permeability of free space, respectively. The three concepts come together here, as each antenna element gets less input power than a single element, resulting in lower E_{p1} and E_{p2} penetrating from each element. They do, however, superimpose to generate a total field E_{pT}, which is dependent on the relative phase of the components. Where there is strong mutual coupling between components, changing the distribution can further reduce E_{pT} and therefore SAR [22]:

$$SAR = \frac{\rho \times |E_{pT}|^2}{M_d} \quad \text{(W/kg).} \tag{5.4}$$

The SAR must adhere to both national and international EMF exposure limits, which are 1.6 W/kg averaged over 1 g of body tissue from the Federal Communications Commission (FCC) [23] and up to 2.0 W/kg averaged over 10 g of body tissue from the International Commission on Non-ionizing Radiation Protection (ICNIRP) [1]. These values are based on the highest transmit power available from the UE [24]. The next two sections go through comprehensive simulations to see how the three concepts presented here affect SAR reduction well within the acceptable limitations.

5.3 Coupling Manipulation Simulation Campaign

In this work, an PIFA, as represented in Figure 5.1, is chosen as a possible antenna that uses the ground plane and has also been employed in earlier studies to minimize SAR [13] while allowing for adequately strong coupling between components. Following the dimensions of a common smartphone, a ground plane with length $L_g = 138$ mm, width $W_p = 68$ mm, and thickness $t = 0.8$ mm was chosen. The PIFA elements from [26] have a fixed width $W_p = 10$ mm and a length L_p that is dependent on the element separation d so that S_{11} (equal to S_{22}) is minimal at 2.4 GHz. Table 5.1 displays discrete values of d ranging from 7 to 62 mm, as well as corresponding values of L_p. Each PIFA element is linked to the ground plane through a connecting pin and a feeding port (as indicated on the right-hand side of Figure 5.1), and the height of each PIFA element above the ground plane is $H_g = 6$ mm. In Figure 5.2, the values of S_{11} and S_{21} (equivalent to S_{12}) are shown for three usage cases: (i) open space; (ii) when using a spherical head homogeneous model; and (iii) when using a more realistic voxel (Donna: accessible in

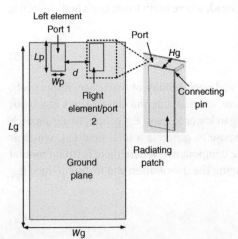

Figure 5.1 A traditional PIFA illustration arranged in a two-element MIMO layout. Source: Jamshed et al. [25]/with permission of IEEE.

Table 5.1 Adjusted values of L_p for each d.

d(mm)	L_p(mm)	d(mm)	L_p(mm)	d(mm)	L_p(mm)
7	21.1	22	21.38	42	20.80
8	21.05	27	21.25	47	20.30
12	21	32	21.25	60	20.90
17	21.22	37	20.80	62	20.90

Source: Jamshed et al. [25]/with permission of IEEE.

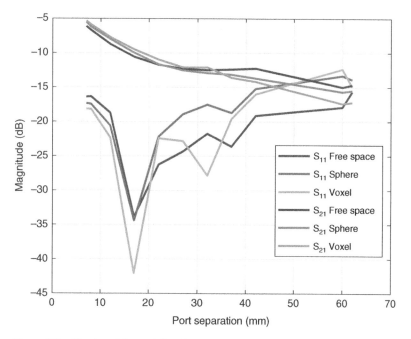

Figure 5.2 Simulated S_{21} and S_{11} of a two-element MIMO PIFA arrangement, positioned with the spherical and voxel based, head models, and in free space, by changing d and keeping resonance frequency fixed at 2.4 GHz. Source: Jamshed et al. [25]/with permission of IEEE.

the CST Voxel family) heterogeneous model, as seen in Figure 5.3. The findings in Figure 5.2 demonstrate that S_{11} is suitably low below −10 dB regardless of d, although S_{21} declines as d increases, as predicted.

The simulation configuration in Figure 5.3 is based on IEEE guidelines [27], where the default handset orientation relative to the head is determined by the voxel model's cheek location. The center of the top of the ground plane, which corresponds to the earpiece, is likewise aligned with the ear. The ground

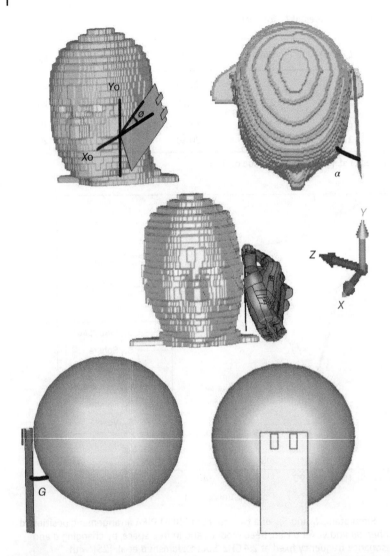

Figure 5.3 Following the IEEE guidelines and aligning UE with the homogeneous spherical and heterogeneous voxel head models used for SAR analysis. Source: Jamshed et al. [25]/with permission of IEEE.

plane is at a tangent with the handset earpiece aligned with the center in the solid sphere model (conductivity 1.42 S/m and dielectric constant 39.9). The G distance between the voxel/sphere and the UE is fixed to 5 mm, which is in accordance with industry norms. The IEEE/IEC 62704-1 method [28] is used in

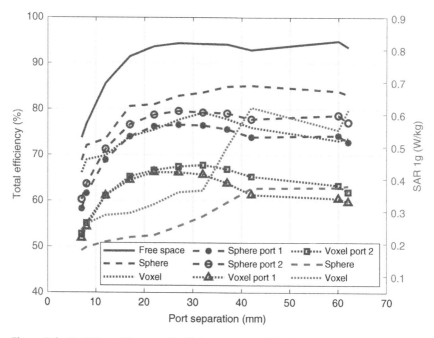

Figure 5.4 Variation of the overall efficiency and the SAR for a sphere with a fixed phase of 0° and voxel head models with a fixed phase of 30° (when *d* is adjusted from 7 to 32 mm) and 0° (when *d* is varied from 37 to 62 mm). It should be noted that the phase angles chosen for the sphere and voxel head models match to the minimal value of the SAR. Source: Jamshed et al. [25]/with permission of IEEE.

both circumstances to determine the maximum averaged SAR (for 1 g of body mass) value by utilizing the maximum long term evolution (LTE) uplink transmit power of 23 dBm, which is spread evenly between the two PIFA components. The influence of hand grip, as seen in Figure 5.3, is also investigated further in Section 5.5. Furthermore, *α* denotes the handset's tilt angle away from the cheek, which is considered to be zero unless when its influence is examined in Section 5.5. The angle of inclination *ϕ* is set at 60°.

Figure 5.4 displays the differences in overall efficiency as a function of *d* on left side of the graph, as well as individual port efficiencies. This is taken into account while the handset is in free space, as well as when the sphere and voxel models are present. Because both ports have similar efficiency in open space, only the efficiency of one port is displayed. When *d* is less than 17 mm, which corresponds to less than 0.14λ, the overall efficiency for both ports begins to deteriorate significantly, with S_{21} ranging from −9 to −11 dB (shown in Figure 5.2). The right side curves represent the improved SAR as a function of *d* for the sphere and voxel models. The phase between elements is adjusted to correspond to the

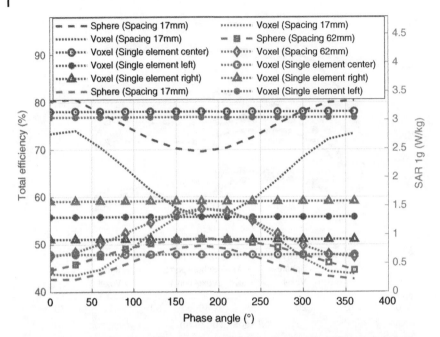

Figure 5.5 SAR and overall efficiency are varied by adjusting the difference of phase between antenna elements, as well as several single element configurations of PIFA (center, right, and left), which are aligned with the spherical and voxel head models. Source: Jamshed et al. [25]/with permission of IEEE.

lowest SAR for each value of d chosen, which is 30° for $d = 7$ to 32 mm and 0° for $d = 37$ to 62 mm. These phase angles are also used by the corresponding efficiency ratings. The results show that when $d = 17$ mm, the SAR may be cut in half when compared with the maximum spacing at 62 mm that can be presumed to be a normal MIMO antenna arrangement. Furthermore, efficiency at 17 mm remains close to that at 62 mm. Efficiency declines significantly at d 17 mm, whereas any additional reduction in SAR is minimal.

Figure 5.5 extends the results of Figure 5.4 by displaying changes in overall antenna efficiency and SAR as a function of the relative phase between the two antenna elements, or ports, when d is set at either 17 or 62 mm, or when only one port is employed. It is obvious that for both sphere and voxel models, the SAR varies considerably with relative phase, as demonstrated in [14]. As shown in Figure 5.4, the SAR is lower for $d = 17$ mm than for $d = 62$ mm; this further indicates that increased coupling between the antenna elements helps to reduce the SAR to a minimal value of 0.29 and 0.21 W/kg for the voxel and sphere models, respectively, at a phase angle of 30°. When $d = 17$ mm, the SAR achieves a maximum value of 1.4 W/kg for the voxel and 0.81 W/kg for the sphere at roughly 180°. It should be noted that the maxima (and minima) of the voxel and sphere models occur at the identical phase angles, and so the composition of the cranial tissues has no effect on this. Furthermore, because the same transmit power is distributed between the two components, the maximum SAR of 1.4 W/kg (for the voxel model) is smaller than the SAR of the individual antenna element whether positioned to the right or left of a phone. The right and left pieces are labeled in Figure 5.1, and their SAR differ because they are positioned differently in relation to the head tissues. Though impractical, the SAR from a single PIFA in the center of the handset, where the earpiece would be, is also plotted due to higher ear absorption, but the voxel result for 17 mm spacing at 30° still reduces SAR by 50%, while compared with a right and left hand element, the reduction is 81% and 90%, respectively. As a result, center spaced antenna components are the greatest solution for reducing SAR in the talk position. When comparing the voxel results at 30° for 17 and 62 mm, the decrease is similarly 50%, as shown in Figure 5.4, which validates employing the greater coupling while maintaining efficiency. As a result, rather than producing heat, the antennas radiate away from the head.

5.4   SAR Analysis and Surface Current

After determining the optimal phase angle for minimizing SAR, this part examines the three physical concepts stated in Section 5.2. The following PIFA configurations are investigated using current distributions in Figure 5.6 and SAR analyses in Figures 5.7 (voxel) and 5.8 (sphere):

1. In Figures 5.6a–c and 5.7a–c, a single element PIFA is put in the center, left, and right of the ground plane.
2. A low coupling level scenario ($d = 62$ mm) with phase angles of 0° and 180° (equivalent to the minimum and maximum SAR values as a function of phase in Figure 5.5), in Figures 5.6d and e, 5.7d and e, and 5.8a and b, respectively.
3. A high coupling scenario ($d = 17$ mm) with phase angles of 30° and 180° (equivalent to the minimum and maximum SAR values as a function of the phase in Figure 5.5), in Figures 5.6f and g, 5.7f and g, and 5.8c and d, respectively.

For clarity, the surface currents in Figure 5.6 are adjusted to a maximum peak colorbar level of 15 A/m, whereas the maximum real current is reported on each sub figure. The current distribution is depicted on the front (top portion of each

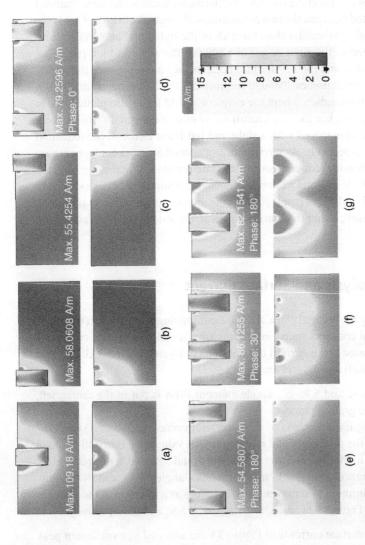

Figure 5.6 Current distribution comparison of various PIFA configurations aligned with the voxel model; (a) A single PIFA element is positioned in the center of the ground plane, (b) A single PIFA element is positioned on the left of the ground plane, (c) A single PIFA element is positioned on the right of the ground plane, (d) two-element PIFA with low coupling arrangement having a 0° difference of phase, (e) two-element PIFA with low coupling arrangement having a 30° difference of phase, (f) two-element PIFA with high coupling arrangement having a 180° difference of phase, (g) two-element PIFA with high coupling arrangement having a 180° difference of phase. Note: Although the maximum current in the colorbar is limited to 15 A/m, the actual maximum current is stated on each subfigure. Source: Jamshed et al. [25]/with permission of IEEE.

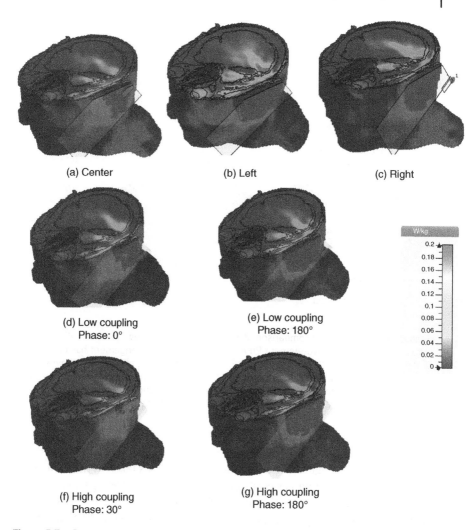

Figure 5.7 Cross section comparison of SAR of various PIFA configurations aligned with voxel model; (a) A single PIFA element is positioned in the center of the ground plane, (b) A single PIFA element is positioned on the left of the ground plane, (c) A single PIFA element is positioned on the right of the ground plane, (d) two-element PIFA with low coupling arrangement having a 0° difference of phase, (e) two-element PIFA with low coupling arrangement having a 180° difference of phase, (f) two-element PIFA with high coupling arrangement having a 30° difference of phase, (g) two-element PIFA with high coupling arrangement having a 180° difference of phase. Source: Jamshed et al. [25]/with permission of IEEE.

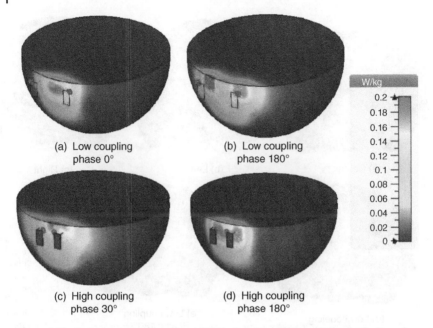

Figure 5.8 Cross section comparison of SAR of various PIFA configurations aligned with sphere model; (a) two-element PIFA with low coupling arrangement having a 0° difference of phase, (b) two-element PIFA with low coupling arrangement having a 180° difference of phase, (c) two-element PIFA with high coupling arrangement having a 30° difference of phase, (d) two-element PIFA with high coupling arrangement having a 180° difference of phase. Source: Jamshed et al. [25]/with permission of IEEE.

subfigure) and back (bottom part of each subfigure) of the UE in the region of the PIFA components. Figures 5.7 and 5.8 are cross sections taken at the level of the earpiece, and the colorbar has been adjusted to a limit of 0.2 W/kg for clarity.

Figure 5.6a–c demonstrates that the feed forms a strong current in each radiating element, and the ground plane has a strong current near to the PIFA. When the PIFA is in the center, a greater spread of current is detected on the ground plane because it interacts differently with the ground plane, while the ear also lowers the SAR through to the head when comparing Figure 5.7a–c. It is worth noting that the left antenna is closer to the head, resulting in a greater SAR than the right antenna (as illustrated in Figure 5.5).

In the low coupling situation shown in Figure 5.8a, where the phase is selected for the lowest SAR, the penetrating fields sum in phase at the midpoint between the two antenna elements, but out of phase at each element. Setting the relative phase in Figure 5.8b from 0° to 180° produces the predicted inverse effect. As a result of evenly dividing the power between the two parts, the lowest SAR happens with the phase chosen so that the penetrating fields with the lowest magnitude add constructively. Increased coupling with closer elements in Figure 5.8c and d, the same concept of penetrating fields accumulating together in and out of phase happens, but there is also a significant shift in the current distribution, particularly on the ground plane of the UE, as seen in Figure 5.6f and g. This will result in a significant shift in the total tangential magnetic field H_{tpT} and penetrating electric field E_{pT}, which will be minimized with a phase of 30°. As a result, the three concepts outlined in Section 5.2 can work together to reduce the SAR.

The sphere model is a valuable tool for analyzing how the physical principles function to get the least SAR for the PIFA components with a spacing d of 17 mm and a relative phase angle of 30°. The similar result happens in a voxel head, but with less clarity, as seen in Figure 5.7d–g, because the existence of the ear causes the penetrating fields to spread unevenly. Nonetheless, the SAR in the ear is demonstrated in Figure 5.7f to be fair. Finally, as with PIFAs, a considerable ground plane is necessary in order to have enough conducting volume to significantly modify the current distribution when the components are closely linked. Smaller conductive structures, such as wire antennas, could not do this.

5.5 Resilience to Different Head Use Cases

Only the fixed position of a 60° inclination angle has been studied up to this point. However, it is also necessary to examine the fluctuations in the SAR when the handset deviates from this position, most notably when the earpiece is offset relative to the ear hole in talk position and/or a change in inclination or tilt angle. Additionally, the consequences of the hand grip in a typical use scenario must be considered. When $d = 17$ mm, the following simulation settings are used to examine the impact on the SAR:

1. The inclination angle ϕ is set to 0°, 30°, 60°, or 90° to reflect the range of angles with which the handset may be held.
2. The tilt angle α is increased to 15°, following the IEEE recommendations [27].
3. By moving the handset in the x or y directions, the UE is offset. More specifically, in Figure 5.3, the value of X_0 or Y_0 is varied by ± 10 mm to achieve four different offsets, namely 10 mm upward, 10 mm downward, 10 mm right, and 10 mm left, to reflect the maximum expected position offset of the handset with regard to the earpiece aligning with the ear hole.

The changes of the SAR as a function of the phase angle for the different aforementioned setups are shown in Figures 5.9 and 5.10. In terms of the inclination angle, the minimal SAR is maintained at a phase angle of 30°, except when ϕ is 90°. Even in this case, it only rises to 0.5 W/kg, which is still significantly lower by

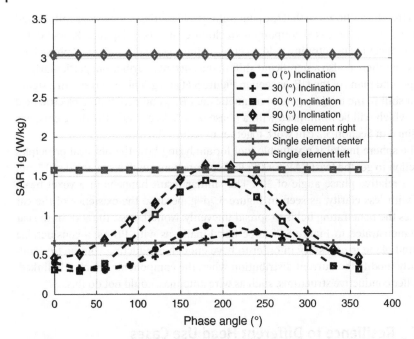

Figure 5.9 SAR vs. phase angle variations between elements of PIFA for a fixed *d* of 17 mm and various angles of inclination compared with single elements at 60° inclination. Source: Jamshed et al. [25]/with permission of IEEE.

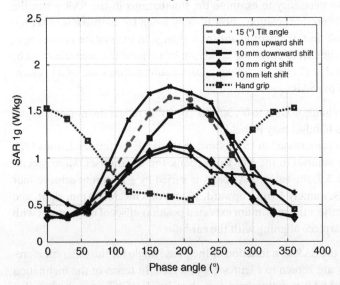

Figure 5.10 The variation of SAR vs. phase angle between elements with fixed element spacing of 17 mm and 60° inclination angle with a change in tilt angle to 15° and spatial earpiece offset in the upward, downward, left, and right directions. Source: Jamshed et al. [25]/with permission of IEEE.

a factor of three or more when compared with a single element on the handset's left or right side. In terms of UE offset, the minimal SAR is still maintained at a phase angle of 30°, even when a 10 mm upward shift is applied. However, its SAR is still only 0.5 W/kg, which is the same as the situation for $\phi = 90°$.

Regarding the influence of hand grip, the worst case scenario depicted in Figure 5.3 is used here, in which the fingers cover the antennae on the phone. The minimum SAR is no longer maintained at a phase angle of 30°, as shown in Figure 5.10, and the curve is practically reversed. In this case, the averaged SAR is computed with the addition of the hand. Figure 5.11a and b clearly show that

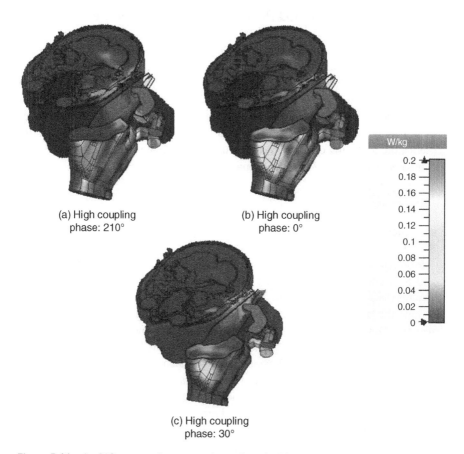

(a) High coupling
phase: 210°

(b) High coupling
phase: 0°

(c) High coupling
phase: 30°

W/kg

0.2
0.18
0.16
0.14
0.12
0.1
0.08
0.06
0.04
0.02
0

Figure 5.11 An SAR penetration comparison aligned with voxel model using a hand grip (a) two-element PIFA with high coupling arrangement having a 210° difference of phase, (b) two-element PIFA with high coupling arrangement having a 0° difference of phase, (c) two-element PIFA with high coupling arrangement having a 30° difference of phase. Source: Jamshed et al. [25]/with permission of IEEE.

significant absorption occurs immediately in the hand. Figure 5.11a and b shows the SAR penetration for relative phase angles of 210° (lowest SAR in Figure 5.10) and 0° (maximum SAR in Figure 5.10). According to Figure 5.10, where the mean SAR is lowest at 210° (according to Figure 5.10), there is actually greater SAR in the skull for this relative phase than for 0°. As a result, the SAR in the skull is still minimal with a phase angle of 30°, as shown in Figure 5.11c, which shows less than 0.08 W/kg at any place in the head. As a result, hand grasp has no negative effect in terms of reducing SAR to the head.

5.6 Analysis of MIMO Performance in Data Mode

A smartphone's accelerometer may identify whether a device is operating away from the head in "data mode," in which case it can be reconfigured to use MIMO precoding rather than a fixed phase for low SAR against the head. However, the capacity can be reduced since the reciprocal coupling should ideally be less than −15 dB [26]. The correlation coefficient has a direct effect on capacity [29], where for a two-element MIMO channel with a sufficiently high signal-to-noise ratio (SNR), the following expression can be used to calculate the capacity loss when compared with perfect MIMO antennas [30]:

$$\overline{C}_{\text{loss}} = -\log_2 \begin{vmatrix} \rho_{1,1} & \rho_{1,2} \\ \rho_{2,1} & \rho_{2,2} \end{vmatrix} \tag{5.5}$$

such that considering S-parameters $\rho_{i,i} = (1 - |S_{i,i}|^2 + |S_{i,j}|^2)$ and $\rho_{i,j} = -(S_{i,i}^* S_{i,j} + S_{j,i}^* S_{j,j})$, i and $j \leq 2$.

Prototypes for the PIFA with d of 17 and 62 mm are shown in Figure 5.12, and S-parameter observations are compared with simulations in Figure 5.13. The results are in good agreement; however, in the case of strong coupling, $|S_{21}|$ reveals the measured result to be −9 dB rather than −11 dB in the simulated scenario. However, there is coupling between the ports that may be used to justify such a discrepancy at such tight spacing. Another crucial measure for closely spaced antennas is that the efficiency must not decline, as shown in Table 5.2 at 2.4 GHz. The predicted and observed peak gains in the elevation plane normal to the ground plane are compared. The results are within 1.6 dB of each other due to the impacts of the connections and measuring cable, but they confirm that efficiency is appropriately maintained. It is obvious that where S_{21} is up to −9 dB, a minor capacity loss must be permitted. Nonetheless, this problem might be solved by installing PIFA antennas with $d = 17$ mm near the earpiece and those

Figure 5.12 Prototype of the UE with (a) $d = 17$ mm and (b) $d = 62$ mm. Source: Jamshed et al. [25]/with permission of IEEE.

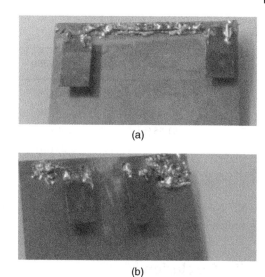

(a)

(b)

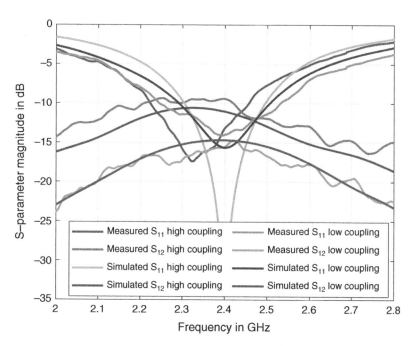

Figure 5.13 Comparison of simulated and measured S-parameters for the UE with $d = 17$ mm (high) and $d = 62$ mm (low) coupling in free space. Source: Jamshed et al. [25]/with permission of IEEE.

Table 5.2 Comparison of UT metrics with $d = 17$ mm and $d = 62$ mm at 2.4 GHz. Simulated (S) and measured (M) results shown.

d (mm)	Efficiency (%)	Gain (dBi)	S_{21} (dB)	$\overline{C}_{loss \, (bits/s/Hz)}$
62	93.38 (S)	5.60 (S)	−15.00 (S)	0.14 (S)
		4.04 (M)	−15.60 (M)	0.20 (M)
17	91.55 (S)	2.56 (S)	−11.10 (S)	0.23 (S)
		2.28 (M)	−9.40 (M)	0.41 (M)

Source: Jamshed et al. [25]/with permission of IEEE.

with $d = 62$ mm at the bottom near the microphone and switching antennas based on the use case.

5.7 Conclusion

In this chapter, a systematic study of how SAR to the human head may be decreased from a smartphone device in talk position utilizing two antenna elements by spreading the source power two ways and regulating the relative phase between them to enable the penetrated fields to superimpose to produce the lowest magnitude has been performed. When tuned, SAR can be decreased by a factor of three or more when compared with a single element. This has been demonstrated to operate consistently independent of the head it is against or the user handling that is used as long as the earpiece is held against the ear. When the device is removed from the body, it can still function as a two-element MIMO antenna, and adaptive approaches can be used.

References

1 ICNIRP (2020). Guidelines for limiting exposure to electromagnetic fields (100 kHz to 300 GHz). *Health Physics* 118 (5): 483–524.

2 Imran, M.A., Héliot, F., and Sambo, Y.A. (2019). *Low Electromagnetic Emission Wireless Network Technologies: 5G and Beyond*. Institution of Engineering and Technology.

3 G.S.38.151 (2020). NR; User Equipment (UE) Multiple Input Multiple Output (MIMO) Over-the-Air (OTA) Performance Requirements. *Release 17*.

4 I.P802.11 (2020). *Telecommunications and Information Exchange Between Systems Local and Metropolitan Area Networks, Part 11: Wireless LAN Medium Access Control (MAC) and Physical Layer (PHY) Specifications*. Advanced Technical Ceramics. Monolithic Ceramics. Gerneral and Textural Properties.

5 Jiang, W., Cui, Y., Liu, B. et al. (2019). A dual-band MIMO antenna with enhanced isolation for 5G smartphone applications. *IEEE Access* 7: 112554–112563.

6 Yuan, X.-T., He, W., Hong, K.-D. et al. (2020). Ultra-wideband MIMO antenna system with high element-isolation for 5G smartphone application. *IEEE Access* 8: 56281–56289.

7 Wang, J., Fujiwara, O., and Takagi, T. (1999). Effects of ferrite sheet attachment to portable telephone in reducing electromagnetic absorption in human head. *IEEE International Symposium on Electromagnetic Compatibility* 2: 822–825.

8 Kitra, M.I., Panagamuwa, C.J., McEvoy, P. et al. (2007). Low SAR ferrite handset antenna design. *IEEE Transactions on Antennas and Propagation* 55: 1155–1164.

9 Abedin, M.F. and Ali, M. (2003). Modifying the ground plane and its effect on planar inverted-F antennas (PIFAs) for mobile phone handsets. *IEEE Antennas and Wireless Propagation Letters* 2: 226–229.

10 Zhang, H.H., Yu, G.G., Liu, Y. et al. (2021). Design of low-SAR mobile phone antenna: theory and applications. *IEEE Transactions on Antennas and Propagation* 69 (2): 698–707.

11 Hwang, J. and Chen, F. (2006). Reduction of the peak SAR in the human head with metamaterials. *IEEE Transactions on Antennas and Propagation* 54: 3763–3770.

12 Gómez-Villanueva, R., Jardón-Aguilar, H., and y Miranda, R.L. (2010). State of the art methods for low SAR antenna implementation. *Proceedings of the 4th European Conference on Antennas and Propagation*, pp. 1–4.

13 Chim, K.-C., Chan, K.C., and Murch, R.D. (2004). Investigating the impact of smart antennas on SAR. *IEEE Transactions on Antennas and Propagation* 52 (5): 1370–1374.

14 Hochwald, B.M., Love, D.J., Yan, S., and Jin, J. (2013). SAR codes. *Information Theory and Applications Workshop (ITA), 2013*, pp. 1–9, IEEE.

15 Baldauf, M.A., Pontes, J.A., Timmermann, J., and Wiesbeck, W. (2007). Mobile MIMO phones and their human exposure to electromagnetic fields. *International Conference on Electromagnetics in Advanced Applications, 2007. ICEAA 2007*, pp. 9–12, IEEE.

16 Li, H., Tsiaras, A., and Lau, B.K. (2017). Analysis and estimation of MIMO-SAR for multi-antenna mobile handsets. *IEEE Transactions on Antennas and Propagation* 65 (3): 1522–1527.

17 Zhao, K., Zhang, S., Ying, Z. et al. (2013). SAR study of different MIMO antenna designs for LTE application in smart mobile handsets. *IEEE Transactions on Antennas and Propagation* 61: 3270–3279.

18 Héliot, F., Jamshed, M.A., and Brown, T.W. (2020). Exposure modelling and minimization for multi-antenna communication systems. *2020 IEEE 91st Vehicular Technology Conference (VTC2020-Spring)*, pp. 1–6, IEEE.

19 Balanis, C.A. (2016). *Antenna Theory: Analysis and Design*. Wiley.

20 Kim, Y.-D., Yang, S.-J., Kang, Y.-S. et al. (2019). Mutual admittance of two arbitrary antennas in nonplanar skew positions based on infinitesimal dipole modeling. *IEEE Transactions on Antennas and Propagation* 67 (11): 6705–6713.

21 Wang, H. (2020). Analysis of electromagnetic energy absorption in the human body for mobile terminals. *IEEE Open Journal of Antennas and Propagation* 1: 113–117.

22 Jamshed, M.A., Heliot, F., and Brown, T. (2019). A survey on electromagnetic risk assessment and evaluation mechanism for future wireless communication systems. *IEEE Journal of Electromagnetics, RF and Microwaves in Medicine and Biology*.

23 Radiofrequency Electromagnetic Fields (1997). Evaluating compliance with FCC guidelines for human exposure to radio frequency electromagnetic fields. *OET Bulletin* 65: 1–53.

24 Federal Communications Commission (2014). Specific Absorption Rate (SAR) for cell phones: what it means for you.

25 Jamshed, M.A., Brown, T.W., and Héliot, F. (2022). Dual antenna coupling manipulation for low SAR smartphone terminals in talk position. *IEEE Transactions on Antennas and Propagation* 70 (6): 4299–4306.

26 Chebihi, A., Luxey, C., Diallo, A. et al. (2008). A novel isolation technique for closely spaced PIFAs for UMTS mobile phones. *IEEE Antennas and Wireless Propagation Letters* 7: 665–668.

27 IEEE Std 1528-2013 (2013). *IEEE Recommended Practice for Determining the Peak Spatial-Average Specific Absorption Rate (SAR) in the Human Head from Wireless Communications Devices: Measurement Techniques (Revision of IEEE Std 1528-2003)*, pp. 1–246.

28 IEC/IEEE 62704-1:2017 (2017). *IEC/IEEE International Standard – Determining the peak spatial-average specific absorption rate (SAR) in the human body from wireless communications devices, 30 MHz to 6 GHz - Part 1: General requirements for using the finite-difference time-domain (FDTD) method for SAR calculations*, pp. 1–86.

29 Chae, S.H., Oh, S.-k., and Park, S.-O. (2007). Analysis of mutual coupling, correlations, and TARC in WiBro MIMO array antenna. *IEEE Antennas and Wireless Propagation Letters* 6: 122–125.

30 Shin, H. and Lee, J.H. (2003). Capacity of multiple-antenna fading channels: spatial fading correlation, double scattering, and keyhole. *IEEE Transactions on Information Theory* 49 (10): 2636–2647.

6

MIMO Antennas with Coupling Manipulation for Low SAR Devices

Muhammad Ali Jamshed[1], Tim W.C. Brown[2], and Fabien Héliot[2]

[1]*James Watt School of Engineering, University of Glasgow, Glasgow, UK*
[2]*Institute of Communication Systems (ICS), Home of 5G and 6G Innovation Centre, University of Surrey, Guildford, UK*

6.1 Introduction

In future wireless communication, the multiple-input multiple-output (MIMO) method is seen as a vital technology for boosting spectrum efficiency and channel capacity [1]. The primary method for increasing capacity at the user equipment (UE) is to increase the number of active antenna components, but this is restricted by space limits at the UE [2]. Several research have attempted to multiply active components while maintaining MIMO performance. Attempts to increase the number of active components while preserving MIMO performance have been undertaken in several studies. The [2], for example, presents an eight-element MIMO antenna for metal-rimmed 5G based mobile phones using a combine architecture of loop and slot antenna. In [3], an eight element MIMO rim antenna for 5G smartphones is given, which integrates the dual antenna pair and orthogonal modes to offer the required envelope cross correlation (ECC). These and other research projects just concentrate on using various methods, such as neutralization lines ([4, 5], etc.), to multiply the number of active antennas while still fulfilling "MIMO" requirements; however, none of them have concentrated on the change in "EM field (EMF)" associated to the human head, which is measured by "specific absorption rate (SAR)" [6].

There are several methods for reducing SAR, such as metamaterials [7], ferrites [8, 9], and so on, but the size constraint at the UE and the complexity of these designs make it practically difficult to maintain the SAR under the given limitations. Previous work on building antennas for UE does not try to control/reduce

Low Electromagnetic Field Exposure Wireless Devices: Fundamentals and Recent Advances, First Edition.
Edited by Masood Ur Rehman and Muhammad Ali Jamshed.

the SAR during operation and merely shows that the SAR adheres to safety limitations. The low SAR values given in [10, 11] are related to the positioning of the single rim antenna element away from the ear, and avoid considering the MIMO arrangement while performing the SAR estimations. Similarly, the authors of [12] give no data pertaining to SAR values. Furthermore, by using our recommended defective ground structure (DGS) enabled design, any two-element MIMO rim antenna may achieve up to a 30% reduction in SAR when compared with their standard value. This chapter provides the first study to investigate the possibility of lowering SAR by placing MIMO antennas integrated into a smartphone's rim.

The remainder of the chapter is structured as follows: The geometry and operation of the suggested antenna are explained in Section 6.2 of the chapter. Section 6.3 discusses the manufacture and modeling of our prototype antenna, as well as the comparison of measured data. Section 6.3 further investigates the fluctuation in the antenna's MIMO performance when the levels of mutual coupling are changed. Section 6.4 confirms the decrease in SAR and investigates the impact of the hand and LCD on antenna efficiency when the amount of coupling is changed. Section 6.5 brings the chapter to a conclusion.

6.2 Working Principle and Antenna Geometry

Depending on the places where the rim is supplied via two ports on the right and left sides of the UE, the recommended antenna achieves two resonance frequencies of 2.1 and 4.3 GHz. These feeds are assisted by a DGS, which manipulates the ground surface current as well as the characteristic modes and, eventually, the SAR. The working mechanism of the proposed antenna is explained in detail in the following section.

6.2.1 Antenna Dimensions

The geometry and the dimensions of the proposed two-element MIMO rim-based dual-frequency antenna design are illustrated in Figure 6.1. The antenna's chassis is 132 mm by 74 mm, which is standard for every smartphone on the market. As a substrate, single-sided FR-4 (thickness = 0.8 mm and ϵr = 4.4) is utilized. Metal plates with 0.3 mm thickness are utilized to cover the chassis' surrounds. To activate the higher modes of resonance, two 1 mm gaps are slit in the shorter edges [11], whilst two 10 mm gaps are slit in the long edges to adjust the degree of coupling between antenna components. Because the majority of the metallic rim is shorted to the ground plane, bigger holes on the longer sides are required to separate the DGS and the rim. The dimensions of each element, as well as

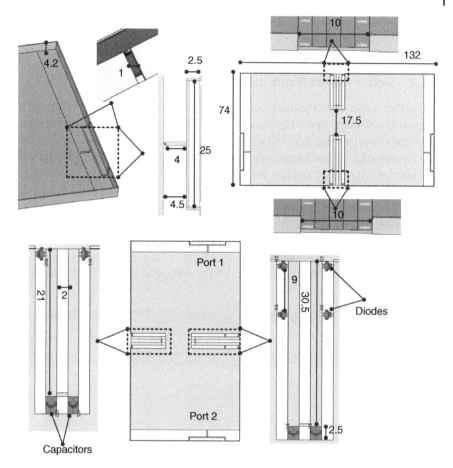

Figure 6.1 The proposed two element MIMO-enabled dual band rim antenna's geometry is shown and discussed using several viewpoints, with all units in millimeters (mm). Source: Jamshed et al. [13]/with permission of IEEE.

the ground clearance, are depicted in Figure 6.1. To avoid an imbalance in the amount of coupling between the two antenna elements, the two periodic DGSs are arranged exactly in the center of the two antenna elements and are regulated by diodes (BAR50-02V, Infinenon). To prevent AC and DC signal mixing, two 100 pF capacitors that can block RF signal are used at the edge of each DGS. Due to the longer wavelength at 2.1 GHz, a single diode pair is enough to achieve a variation in the level of coupling, whereas, two identical pairs of the diodes are needed to change coupling levels at 4.3 GHz. Moreover, the dual-band operation of the proposed antenna, the sides of the rim (longer edges) responsible for resonating the antennas at 2.1 and 4.3 GHz, and the working of the DGS are further explained in the preceding sections using the current distribution and the characteristic mode analysis (CMA) analysis.

6.2.2 Surface Current Distribution

To further understand the dual-band functioning and the DGS, we performed a current distribution study. Figures 6.2 and 6.3 show simulated current distribution graphs at operating frequencies (4.3 and 2.1 GHz). Figure 6.2a illustrates that the shorter edge of the longer side of the metallic rim excites the 2.1 GHz, but the shorter edge of the smaller side supports 4.3 GHz (Figure 6.2b). As seen in Figure 6.2a, the folded dipole branch's left side is where the current concentrates most of its energy, creating a resonant mode with a λ/4 frequency that is similar to a monopole structure. As seen in Figure 6.2b, the current is uniformly distributed on both arms of the folded dipole, resulting in a λ/2 resonant mode, similar to a dipole structure. Figure 6.3 shows the variation in surface current when DGSs are switched on or off by altering the level of DC voltage at the diodes. Figure 6.3a

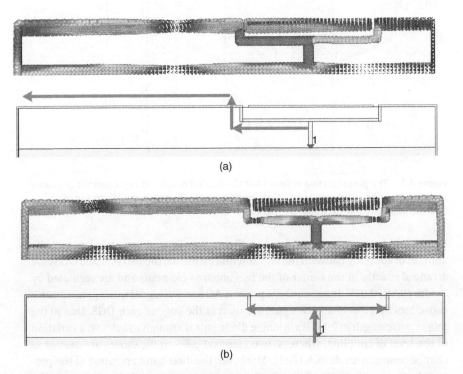

(a)

(b)

Figure 6.2 Surface current plots simulated (a) 2.1 GHz (b) 4.3 GHz. Source: Jamshed et al. [13]/with permission of IEEE.

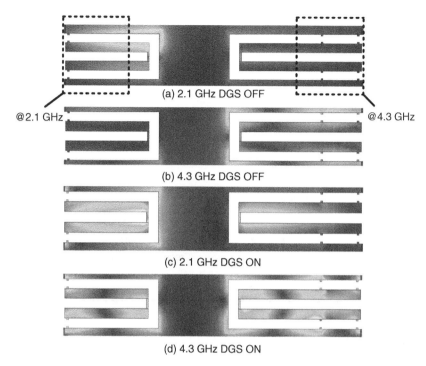

@2.1 GHz

@4.3 GHz

(a) 2.1 GHz DGS OFF

(b) 4.3 GHz DGS OFF

(c) 2.1 GHz DGS ON

(d) 4.3 GHz DGS ON

Figure 6.3 Surface current graphs simulated at 2.1 GHz and 4.3 GHz for scenarios where DGS is either ON or OFF. (a) 2.1 GHz DGS OFF, (b) 4.3 GHz DGS OFF, (c) 2.1 GHz DGS ON, (d) 4.3 GHz DGS ON. Source: Jamshed et al. [13]/with permission of IEEE.

and c shows surface current variations when the antenna connection is adjusted at 2.1 GHz. The current levels fluctuate substantially when the DGS is activated or deactivated, suggesting that the degree of antenna coupling can be successfully adjusted using the DGS. A similar discovery might be made concerning the DGS, which regulates the coupling values at 4.3 GHz.

6.2.3 Frequency Region Analysis

In this chapter, we employ the modal significance (MS), which is based on the theory of characteristic mode (TCM) [14], to get insight into the electromagnetic characteristics of our fractured metal rim construction. The MS $n = 1$ suggests that a nth characteristic mode (CM) might theoretically resonate [15] at a given frequency. We used CM on a fractured metal rim frame to test the dual-band performance of our proposed antenna design by signaling probable CMs at the required operating frequencies. The current modes of a shattered metal rim construction with a ground plane are determined in Figure 6.4a by applying

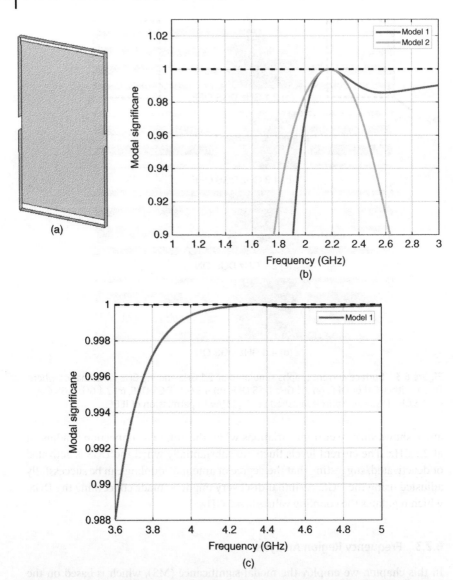

Figure 6.4 An example of a CMA study of a shattered metallic rim. (a) Ground with broken metal RIM model for CMA, (b) Modal significance at lower band, and (c) Modal significance at hiigher band. Source: Jamshed et al. [13]/with permission of IEEE.

the integral equation solver in CST. The overall thickness of the construction is 0 mm, and it has the attributes of a perfect electric conductor (PEC). The antenna construction does not include a feeding port for CMA operation. Figure 6.4b shows that Mode 1 and Mode 2 resonate at 2.1 GHz with MS = 1, whereas Mode 1 can resonate at 4.3 GHz in Figure 6.4c. There are several techniques to trigger these resonance modes; in this chapter, a folded dipole is utilized.

6.3 Antenna Measurements

We created a mock-up prototype of our antenna design to validate it through experimental findings, as shown in Figure 6.5. To prevent fabrication mistakes, two versions of it have been fabricated: one replicating the DGS ON arrangement (see Figure 6.5a), which employs passive connections to activate the DGS, and the other one is DGS OFF arrangement (see Figure 6.5b). The simulated S-parameters for both arrangements, in terms of frequency, are compared with measured results and are shown in Figure 6.6. Because our antenna construction is symmetric, only the results of a single port are displayed here. The measured and simulated S-parameter findings correspond well and reveal a shift in coupling levels when the DGS is either deactivated or activated. Based on simulation data, the level of coupling at 2.1 GHz varies between −11 and −14 dB, and at 4.3 GHz between −13 and −18 dB, depending on whether the DGS is engaged or off. The calculated and observed normalized radiation patterns for the four instances, 2.1 GHz DGS ON and OFF, 4.3 GHz DGS ON and OFF, are compared in Figure 6.7. For both the DGS ON and OFF situations, a nearly omnidirectional radiation pattern is found at the two frequencies. These dipole-like radiation patterns are appropriate for practical communication. Changing the strength of antenna coupling had little effect on the radiation patterns, except for a slight fluctuation in the reported radiation pattern at 2.1 GHz DGS OFF, which is likely attributable to measurement campaign imperfections and/or manufacturing faults.

6.3.1 MIMO Performance

The level of near field coupling of an antenna design is important in determining its MIMO performance. It is commonly known that any coupling level less than −15 dB does not affect MIMO performance [16]. However, as shown in Figure 6.6, the amount of coupling level of the antenna can reach −11 dB, making it critical to investigate its influence. The system's spectral efficiency is directly impacted by the ECC, [17]; therefore, it is vital to show its effect in addition to the capacity loss.

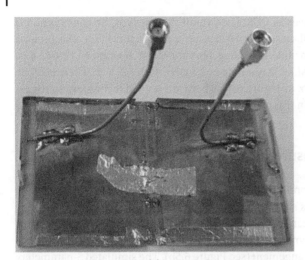

(a)

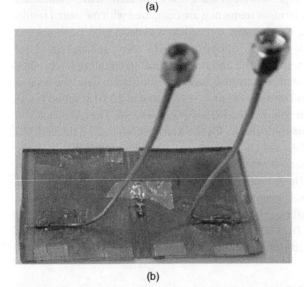

(b)

Figure 6.5 Prototype of (a) DGS ON configuration (b) DGS OFF configuration of the proposed rim antenna design. Source: Jamshed et al. [13]/with permission of IEEE, THE UNIVERSITY OF GLASGOW PRESS.

The mathematical model for ECC and the capacity loss, in terms of S-parameters, is included in the [17]. The measured values of ECC for the DGS OFF case are 0.00017 at 2.1 GHz and 0.00084 at 4.3 GHz, whereas the measured values for the DGS ON case are 0.01 at 2.1 GHz and 0.002 at 4.3 GHz. The measured capacity loss for the DGS OFF case is 0.109 at 4.3 GHz and 0.101 at 2.1 GHz, whereas the capacity loss for the DGS ON case is 0.19 at 4.3 GHz and 0.28 at 2.1 GHz. It should

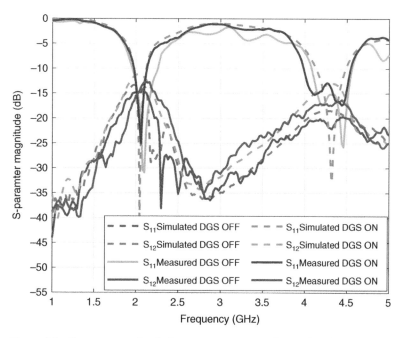

Figure 6.6 S-parameters simulation vs. measurement. Source: Jamshed et al. [13]/with permission of IEEE.

be noted that the ECC and capacity loss estimated using the observed S-parameter are both less than the needed thresholds, 0.4 bits/sec/Hz and 0.05, to operate in MIMO mode. Furthermore, raising the strength of antenna coupling increases the capacity loss and the ECC while decreasing the gap between the estimated and the threshold values.

6.4 Efficiency and SAR Analysis

When the UE comes into close contact with the human body, an increase in mutual coupling can minimize SAR. An in-depth SAR study and positioning the suggested antenna close to the human body in the most often utilized locations proved this. The three locations are depicted in Figure 6.8, namely, putting the handset close to the head, in front, or in side pockets. For the investigation, the Gustav voxel model is employed. The maximum averaged SAR for 10g of body is computed using IEEE/IEC 62704-1 method [18], the 23 dBm transmit power (uplink transmit power for LTE) is spread evenly between the two parts, and they are co-phased.

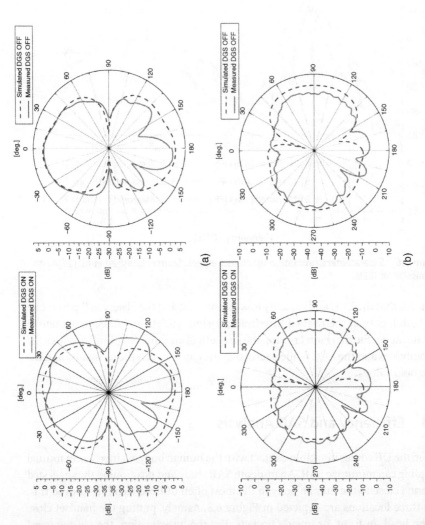

Figure 6.7 Far-field radiation pattern simulation vs. measurement (a) Azimuth (2.1 GHz) (b) Elevation (4.3 GHz). Source: Jamshed et al. [13]/with permission of IEEE.

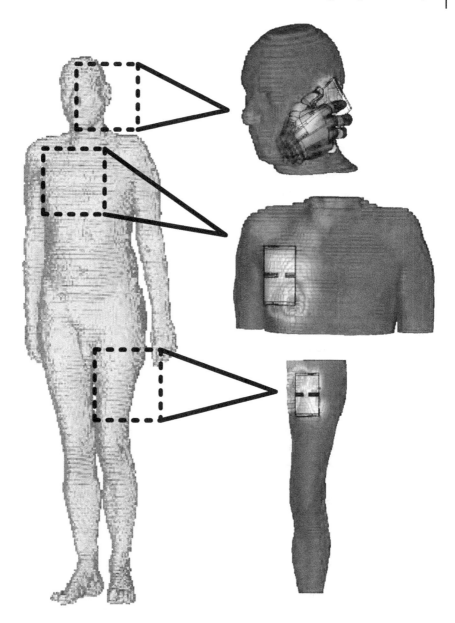

Figure 6.8 An example of a voxel model used to investigate the influence of SAR on modifying the antenna coupling. Source: Jamshed et al. [13]/with permission of IEEE.

Table 6.1 Changes in SAR (10g (W/Kg)) value for scenarios when DGS is either ON or OFF

Resonance frequencies	SAR	Resonance frequencies	SAR	Resonance frequencies	SAR
DGS ON (Talk position)		DGS ON (Front)		DGS ON (Side)	
2.1 GHz	0.8	2.1 GHz	0.1	2.1 GHz	0.3
4.3 GHz	0.9	4.3 GHz	0.1	4.3 GHz	0.2
DGS OFF (Talk position)		DGS OFF (Front)		DGS OFF (Side)	
2.1 GHz	1.2	2.1 GHz	0.1	2.1 GHz	0.4
4.3 GHz	1.1	4.3 GHz	0.1	4.3 GHz	0.2

Source: Jamshed et al. [13]/with permission of IEEE.

Table 6.1 illustrates the fluctuation in SAR values at both 2.1 and 4.3 GHz, and it is obvious that the SAR is lowered when the DGS is activated for each of the employed situations. At 2.1 GHz, for example, when the UE is near to the head, the SAR is effectively decreased by 30%, but at 4.3 GHz, the reduction in SAR is around 14%. This is owing to the fact that when the DGS is enabled, the intensity of antenna coupling increases for both frequencies, as seen in Figure 6.6. Furthermore, as illustrated in Figure 6.3, when the DGS is enabled, the coupling between the antennas rises, resulting in a reduction in the near field coupling from both active antenna element towards the human body, which results in a reduction in SAR.

We have compared our work to that of [10–12] in terms of SAR. The antennas are reproduced, and their SAR values are calculated using a similar setup, i.e. power is evenly split and the antenna elements are co-phased, as used in this chapter, and only for the position, where a cellphone is being used for making voice-calls [19]. The reproduced designs reported in [10–12], estimates SAR to be 1.4 W/kg [10], 1.7 W/kg [11], and 1.1 W/kg [12], respectively. Whereas, our proposed design estimates SAR to be 0.8 W/kg, hence, achieving a 30% reduction in SAR. These results indicate that our suggested design may greatly reduce the SAR.

A rigorous analysis was performed to investigate the variance in overall antenna efficiency when activating or deactivating the DGS in open space while accounting for the LCD and human hand. The simulation configuration for the whole antenna efficiency investigation is shown in Figure 6.9. To simulate the hand effects, the CTIA posable hand model with a $\epsilon r = 1$ spacer is employed. The results for these modifications are shown in Table 6.2, and they indicate that even in the worst-case situation, when the antenna is equipped with an LCD, held with a human hand,

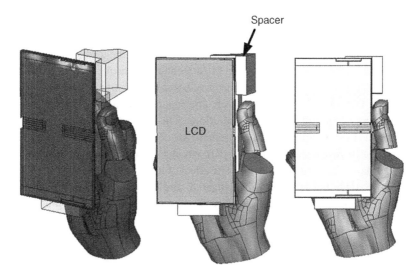

Figure 6.9 Demonstration of the influence of hand and LCD on the antenna design. Source: Jamshed et al. [13]/with permission of IEEE.

Table 6.2 Changes in overall antenna efficiency in (dB)

Antenna elements	Free space	LCD	Hand	LCD with Hand
		DGS Activated		
4.3 GHz(Element 1)	−0.85	−1.32	−1.75	−2.16
2.1 GHz(Element 1)	−0.95	−1.66	−2.02	−2.52
4.3 GHz(Element 2)	−0.85	−1.32	−2.75	−2.79
2.1 GHz(Element 2)	−0.95	−1.66	−2.21	−2.13
		DGS Deactivated		
4.3 GHz(Element 1)	−0.39	−0.85	−1.34	−1.72
2.1 (Element 1)	−0.78	−1.47	−2.04	−2.58
4.3 GHz(Element 2)	−0.39	−0.85	−2.47	−2.43
2.1 GHz(Element 2)	−0.78	−1.47	−2.34	−2.25

Source: Jamshed et al. [13]/with permission of IEEE.

and has a high coupling level, adequate antenna efficiency is still maintained. For example, when operating at 2.1 GHz in open space, there is a 2% difference in overall antenna efficiency while comparing the two DGS OFF and ON cases for element 1.

6.5 Conclusion

The mutual coupling is employed in this chapter to change the level of SAR. The authors present a two element dual-band MIMO-enabled rim antenna design that employs a DGS (periodic version) to effectively boost the amount of coupling levels when the UE comes into close contact. The suggested antenna's operation is validated using CMA and existing distributions. The suggested antenna is built, and the results of simulation and measurement are compared. The measured and simulated results reveal that the MIMO performance is satisfactory, and the SAR analysis validates the suggested design. The effect of hand and LCD on the performance of the suggested antenna is investigated, and the total antenna efficiency indicates an acceptable fluctuation. Finally, the suggested antenna contributes to the creation of a new domain for the design of context-aware UEs.

References

1 Li, W., Lin, W., and Yang, G. (2014). A compact MIMO antenna system design with low correlation from 1710 MHz to 2690 MHz. *Progress in Electromagnetics Research* 144: 59–65.

2 Ren, A. and Liu, Y. (2019). A compact building block with two shared-aperture antennas for eight-antenna MIMO array in metal-rimmed smartphone. *IEEE Transactions on Antennas and Propagation* 67 (10): 6430–6438.

3 Sun, L., Li, Y., Zhang, Z., and Feng, Z. (2019). Wideband 5G MIMO antenna with integrated orthogonal-mode dual-antenna pairs for metal-rimmed smartphones. *IEEE Transactions on Antennas and Propagation* 68 (4): 2494–2503.

4 Ban, Y.-L., Li, C., Wu, G., Wong, K.-L. et al. (2016). 4G/5G multiple antennas for future multi-mode smartphone applications. *IEEE Access* 4: 2981–2988.

5 Wong, K.-L., Tsai, C.-Y., Lu, J.-Y. et al. (2016). Compact eight MIMO antennas for 5G smartphones and their MIMO capacity verification. *2016 URSI Asia-Pacific Radio Science Conference (URSI AP-RASC)*, pp. 1054–1056, IEEE.

6 Jamshed, M.A., Heliot, F., and Brown, T.W. (2019). A survey on electromagnetic risk assessment and evaluation mechanism for future wireless communication systems. *IEEE Journal of Electromagnetics, RF and Microwaves in Medicine and Biology* 4 (1): 24–36.

7 Gómez-Villanueva, R., Jardón-Aguilar, H., and y Miranda, R.L. (2010). State of the art methods for low SAR antenna implementation. *Proceedings of the 4th European Conference on Antennas and Propagation*, pp. 1–4.

8 Kitra, M.I., Panagamuwa, C.J., McEvoy, P. et al. (2007). Low SAR ferrite handset antenna design. *IEEE Transactions on Antennas and Propagation* 55: 1155–1164.

9 Wang, J., Fujiwara, O., and Takagi, T. (1999). Effects of ferrite sheet attachment to portable telephone in reducing electromagnetic absorption in human head. *IEEE International Symposium on Electromagnetic Compatibility* 2: 822–825.

10 Ban, Y.-L., Qiang, Y.-F., Wu, G. et al. (2016). Reconfigurable narrow-frame antenna for LTE/WWAN metal-rimmed smartphone applications. *IET Microwaves, Antennas & Propagation* 10 (10): 1092–1100.

11 Stanley, M., Huang, Y., Wang, H. et al. (2017). A novel reconfigurable metal rim integrated open slot antenna for octa-band smartphone applications. *IEEE Transactions on Antennas and Propagation* 65 (7): 3352–3363.

12 Zhang, L.-W., Ban, Y.-L., Guo, J. et al. (2018). Parallel dual-loop antenna for WWAN/LTE metal-rimmed smartphone. *IEEE Transactions on Antennas and Propagation* 66 (3): 1217–1226.

13 Jamshed, M.A., Brown, T.W., and Héliot, F. (2022). Dual band two element rim based MIMO antennas with coupling manipulation for low SAR mobile handsets. *Progress in Electromagnetics Research.*

14 Garbacz, R. and Turpin, R. (1971). A generalized expansion for radiated and scattered fields. *IEEE transactions on Antennas and Propagation* 19 (3): 348–358.

15 Chen, Y. and Wang, C.-F. (2015). *Characteristic Modes: Theory and Applications in Antenna Engineering.* Wiley.

16 Chen, X., Zhang, S., and Li, Q. (2018). A review of mutual coupling in MIMO systems. *IEEE Access* 6: 24706–24719.

17 Chae, S.H., Oh, S.-k., and Park, S.-O. (2007). Analysis of mutual coupling, correlations, and TARC in WiBro MIMO array antenna. *IEEE Antennas and Wireless Propagation Letters* 6: 122–125.

18 IEC/IEEE 62704-1:2017 (2017). *IEC/IEEE International Standard – Determining the Peak Spatial-Average Specific Absorption Rate (SAR) in the Human Body from Wireless Communications Devices, 30 MHz to 6 GHz - Part 1: General Requirements for Using the Finite-Difference Time-Domain (FDTD) Method for SAR Calculations,* pp. 1–86.

19 IEEE Std 1528-2013 (2013). *IEEE Recommended Practice for Determining the Peak Spatial-Average Specific Absorption Rate (SAR) in the Human Head from Wireless Communications Devices: Measurement Techniques (Revision of IEEE Std 1528-2003),* pp. 1–246.

7

Reinforcement Learning and Device-to-Device Communication for Low EMF Exposure

Ali Nauman[1], Muhammad Ali Jamshed[2], and Sung Won Kim[1]

[1]*Department of Information and Communication Engineering, Yeungnam University, Gyeongsan-si, South Korea*
[2]*James Watt School of Engineering, University of Glasgow, Glasgow, UK*

7.1 Introduction

The International Telecommunication Union (ITU) has put forward the concept of International Mobile Telecommunication-2030 (IMT-2030) network paradigm or Beyond-5G (B5G) network. The new verticals in the IMT-2030 network includes Further enhanced Mobile Broadband (FeMBB), Enhanced Ultra-Reliable and Low Latency Communication (ERLLC), and ultra-massive Machine Type Communication (umMTC) [1]. The applications of these service type verticals are diverse and the requirements are heterogeneous in nature. The common network requirement in these service types are ultra-reliability in terms of Packet Delivery Ratio (PDR) and low latency communication. Almost every parameter of network, that is, throughput, data rate, and energy efficiency, are directly proportional to PDR; whereas, the latency and spectrum resources are inversely proportional to PDR. Therefore, reliable transmission in next generation communication is considered as of utmost importance. Moreover, the concept of Internet-of-Everything (IoE) in IMT-2030 paradigm require massive spectrum access due to exponential increase in connected devices. Currently, 23 billion devices are connected to the Internet, and this number is expected to increase to about 75 billion by 2025 [2]. The demand for Machine Type Communication (MTC) has increased due to the exponential growth of IoT devices, according to citation [3]. The three subcategories of MTC are long-range (range $\geq 100\,\text{m}$), medium-range ($10\,\text{m} < \text{range} < 100\,\text{m}$), and short-range (range $\leq 10\,\text{m}$). IoT devices have a limited amount of energy, processing power, and memory. Low Power Wide Area (LPWA) technology must be standardized for long-range MTC across such constrained devices.

Low Electromagnetic Field Exposure Wireless Devices: Fundamentals and Recent Advances, First Edition.
Edited by Masood Ur Rehman and Muhammad Ali Jamshed.

In Release 13 of Long-Term Evolution-Advanced (LTE-A), the 3rd Generation Partnership Project (3GPP) introduced narrowband-Internet of things NB-IoT [4]. NB-IoT is intended to increase spectrum efficiency and provide comprehensive and extensive coverage [5]. One of the licensed LPWA technologies, NB-IoT offers a transmission range of over 3 km in urban areas and 15 km in open areas, with excellent MTC penetration potential. Supporting IoT devices with a 10-year expected lifespan is the primary goal of NB-IoT. The direct integration of NB-IoT into LTE or GSM networks to share spectrum and repurpose existing gear to save deployment costs is one of the appealing features of this technology. According to ABI research, NB-IoT will cover approximately 60% of the total 3.6 billion LPWA connections by 2026 [6]. For downlink and uplink communication, NB-IoT needs one Physical Resource Block (PRB) of the LTE spectrum, or 180 kHz from the system bandwidth [7]. The limited one PRB bandwidth tightens the resources of the NB-IoT system. The difficulties in deploying NB-IoT become more severe if these inadequate resources are not used effectively.

IoT devices often concentrate on uplink transmission to send the obtained data to the gateway/sink node or the cloud server [8]. One of the top IoT research issues is efficient uplink data transmission. One of the key strategies used in NB-IoT to increase coverage and dependability is the repetition of control and data signals, which involves 128 re-transmissions for uplink and 2048 re-transmissions for downlink. However, when the number of transmission repetitions rises, network efficiency declines [9]. Additionally, deep indoor deployments of NB-IoT devices result in an extra penetration loss of up to 20 dB. Repetition, high transmission power, and longer Transmission Time Interval (TTI) increase time delay and energy consumption, which ultimately raises Electromagnetic Field (EMF) exposure in order to give extended coverage. Therefore, a dependable and effective uplink transmission mechanism is essential for applications requiring widespread deployment and low EMF exposure.

In LTE-A Release 12, the 3GPP standardized Device-to-Device (D2D) communication, also known as Proximity Services (ProSe) [10]. D2D communication has recently drawn a lot more attention due to the limited network resources caused by the rise in linked devices. In Releases 16 and 17, the 3GPP presented more than 24 use cases of D2D communication [11]. One of the D2D use-cases that the 3GPP standardizing committee has authorized is the provision of a reliable mechanism to help the distant User Equipment UE send the collected vital data to the evolved NodeB (eNB) by serving as network relay. The remote UE is either partially covered or not covered at all. The D2D communication uses nearby cellular devices that can function as relay nodes [12]. Network direct and distributed modes of operation for D2D communication are standards. On a remote UE's request in network directed, the eNB controls the choice of the relay UE [11]. While in a distributed system, resources are pooled and communication between the relay UE and remote UE is direct. By lowering the amount of control messages sent out in a distributed system, eliminating the eNB improves performance. One of the enticing features is the incorporation of NB-IoT within the LTE-A standard. As a result, it is projected that NB-IoT-based D2D communication in dispersed systems will improve spectral efficiency and boost wireless networks' performance in terms of data transmission.

Machine Learning (ML) mimics the human brain to enhance its capability for computer vision, image processing, parallel and distributed processing, analytic, and prediction [13]. Reinforcement learning (RL) is a sort of machine learning (ML) in which the learner (agent) does not already know which action to take to maximize the numerical reward (to move in the direction of the main objective). However, the agent must use a hit-and-trial approach to determine which action will produce the most reward. The same problems are optimized using the conclusions drawn from each trial's results. The choice of an RL algorithm is influenced by the issue description and dependent on how well the challenge has been understood. By identifying states, actions, and subsequent rewards, any RL problem can be created. The RL algorithms learn from real-time experience rather than being trained on prior data sets, which makes them less computationally complex than other supervised and unsupervised techniques.

7.1.1 Contribution of Chapter

The goal of the current study was to improve the dependability of UEs that are either outside of the cell's coverage region, undergoing deep fade, or unable to connect to the network due to widespread distribution. Incorporating D2D communication and utilizing NB-integrated IoT's feature within the LTE-A network for D2D communication will be extremely beneficial for the 5G and Beyond-5G networks as a result (B5G). However, the best relay UE selection is crucial to achieving the best end-to-end delivery ratio (EDR). To improve network dependability, the best relay UE choice was heavily subsidized. Therefore, we have defined the relay selection problem as a Markov decision process (MDP) problem at the remote UE and solve it using two-step RL algorithms that, in the first step, examine the available and eligible relay UEs and, in the second, learn the behavior of accomplished EDR. The suggested Intelligent-D2D communication (RL-ID2D) allows for adaptivity in selecting the best relay UE under a variety of relay UE parametric situations, and it delivers the best EDR. As a result, LTE-performance A's is improved in terms of EDR and coverage improvement.

In the meantime, a system-level simulation for narrowband-D2D in LTE-A is created and used to mimic the suggested scenario. The results of simulations and the analysis of performance evaluations demonstrate that the distant UE-based nature of our suggested approach improves network performance. We looked at how RL-based relay selection behaved in two different learning settings. The following are this study's significant contributions:

1. Description and analysis narrowband D2D communication in LTE-A.
2. *Narrowband D2D*: This study adapts the D2D communication as a routing extension for relaying NB-IoT UE data to the eNB in order to maximize the EDR.
3. *OptPRS*: This study formulates the D2D relay selection problem as an optimization problem to maximize the EDR. In order to solve this problem, we propose a simple yet effective solution called the optimum potential relay set (optPRS). The proposed optPRS forms a set of potential relays by comparing the essential parameters with very low complexity.
4. *Intelligent-D2D*: Furthermore, to ensure the maximization of the EDR in dynamic environment, this article models the relay selection as an MDP problem and utilizes RL based ML to solve the problem. The proposed "RL-ID2D" mechanism selects the relay efficiently from the PRS to maximize the EDR and ultimately improves energy efficiency.
5. *Performance Evaluation*: To validate the performance of the proposed intelligent scheme, simulation results are presented: a comparison of RL-ID2D with the pre-existing narrowband-D2D working in network directed system, i.e. deterministic narrowband-D2D [14] and opportunistic narrowband D2D [15] working in distributed decentralized system over the remote UE in an opportunistic manner.

7.1.2 Chapter Organization

The rest of this study is divided into the following sections. The preliminary discussions of D2D communication, NB-IoT, and QL are covered in Section 7.2. Recent investigations in this area are presented in Section 7.3. The system model, problem formulation, and suggested RL-based intelligent D2D communication are covered in Section 7.4 (RL-ID2D). Section 7.5 presents the outcomes of the simulation and a thorough analysis. The last section, Section 7.6, wraps off the chapter.

7.2 Background

With the exponential increase in wireless connected devices to internet in billions, which is expected to increase more abruptly by 2025, there is a dire need of efficient LPWA technology and mechanism to tackle massive deployment and spectrum scarcity. NB-IoT and D2D communications are introduced by 3GPP in Release 13 and 12. NB-IoT is expected to cover 60% of LPWA by 2026 and D2D communication is considered to be a viable solution for coverage enhancement and reliable communication. The amalgamation of ML and D2D is considered an important piece in B5G networks [16]. In this section, we present the main background in which our study is based on.

7.2.1 Narrowband Internet of Things (NB-IoT)

The 3GPP's NB-IoT, which was included in LTE-A Release 13 NB-IoT, is intended to increase spectrum efficiency and provide extensive and comprehensive coverage. One of the permitted LPWA technologies, NB-IoT, offers a transmission range of over 3 km in urban areas and 15 km in open areas, with excellent MTC penetration potential. Supporting IoT devices with a 10-year expected lifespan is the primary goal of NB-IoT. The ability of NB-IoT to integrate with LTE-A or GSM networks to share spectrum and reuse the same hardware to reduce deployment costs is one of the appealing features of this technology [4].

NB-IoT uses one PRB of 180 kHz in the frequency domain for downlink and uplink transmission, which splits into 12 sub-carriers of 15 kHz each. NB-IoT can be deployed directly in the LTE or GSM spectrum in three different modes of operations to scale down the deployment costs. When NB-IoT UE is first powered on, it searches for carrier channel, thus the deployment mode should be clear to the NB-IoT UE [17].

Following are the deployment modes of NB-IoT (as shown in Figure 7.1):

– In-band Mode: One of the PRBs of the LTE spectrum is allocated for NB-IoT deployment. The total power of eNB is shared between LTE and NB-IoT.
– Stand-alone mode. NB-IoT can also be deployed within 200 kHz of the GSM spectrum. NB-IoT can exploit the power of BS, which significantly improves the coverage of the system.
– Guard-band mode. The guard-band of the LTE spectrum is utilized for NB-IoT deployment.

In this study we have considered in-band deployment mode, upon selection of relay UE, it is assumed that one PRB of 180 kHz is allocated for remote UE within relay UE sub-frame.

7.2.1.1 Frame Structure

In order to be compatible with LTE-A, NB-IoT has a frame structure and numeric system comparable with LTE-A. There are 10 sub-frames in each frame of 10 milliseconds. The sub-frame lasts for one millisecond. Seven Orthogonal Frequency Division Multiplexed (OFDM) symbols plus a standard Cyclic Prefix are distributed evenly among two slots each measuring 0.5 milliseconds in duration for each sub-frame (CP). One PRB is equivalent to one slot that has 12 SCs at a frequency of 15 kHz and seven OFDM symbols in the time domain.

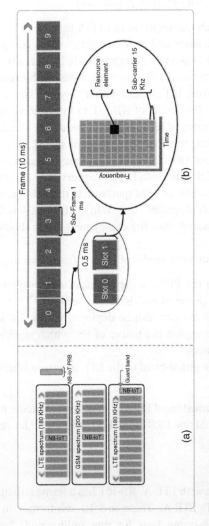

Figure 7.1 (a) NB-IoT deployment modes, (b) NB-IoT frame structure.

Both single-tone and multitoned transmissions are supported via uplink. The single tone occupies 3.75 or 15 kHz Sub-Carrier Spacing (SCS) numerology. The 15 kHz numerology is identical to LTE-A, but the 3.75 kHz numerology uses 2 ms slots. The same 15 kHz SCS is used for multi-tone transmission, which is based on Single Carrier Frequency Division Multiple Access (SC-FDMA). Downlink exclusively employs Quadrature Phase Shift Keying (QPSK), while uplink uses Binary Phase Shift Keying (BPSK) or both. The NB-IoT standard data rate for uplink and downlink transmissions is 160–200 and 160–250 Kbps, respectively. NB-IoT uses 128 re-transmissions for uplink and 2048 re-transmissions for downlink for coverage augmentation [18].

7.2.2 Device-to-Device (D2D) Communication

D2D refers to a direct communication link with nearby devices without considering the intervention of the cellular networks. D2D communication is also defined as Proximity-based Service (ProSe) [19]. D2D was included in the LTE-A Release 12 [20]. The 3GPP standardizing community has approved the proposal of integrating D2D communication into LTE-A. D2D communication has been classified into out-band D2D and in-band D2D communication. In-band, also referred to as LTE Direct, uses a licensed spectrum while out-band D2D exploits an unlicensed spectrum of other wireless enabling technologies that support D2D communication such as IEEE 802.11 (WiFi-Direct) or IEEE 802.15 (Bluetooth) [20]. The D2D UE can access the licensed spectrum in shared mode (also refer as non-orthogonal/underlay mode) or dedicated mode (also known as orthogonal/ overlay mode). The use of D2D communication leads to multiple advantages such as high packet delivery rate, minimum delay, better spectrum re-usability, and low energy consumption. The 3GPP has defined more than 24 use-cases of D2D communication in Release 16 and 17. Few of them are shown in the Figure 7.2.

One of the use-case of D2D communication define by 3GPP is to relaying the data of remote UE. The relaying application of D2D communication is further divided into three scenarios, i.e. in-coverage, out-of-coverage, and partial coverage. In-coverage contain both remote UE and relay UE are within the coverage of the network. In out-of-coverage both the remote and relay UE are out of coverage and establish direct link, while the partial coverage scenario remote UE is out the coverage area of the network and relay UE is within the network coverage as shown in the Figure 7.3. In this study, the in-coverage and partial coverage are considered, where a remote UE that is experience deep fade, congestion, or coverage unavailability need to connect to a relay UE for uplink transmission [21]. The D2D communication is standardized to be used in two deployment modes, i.e. network directed and distributed direct communication. In network directed communication, based on the scenario where the devices

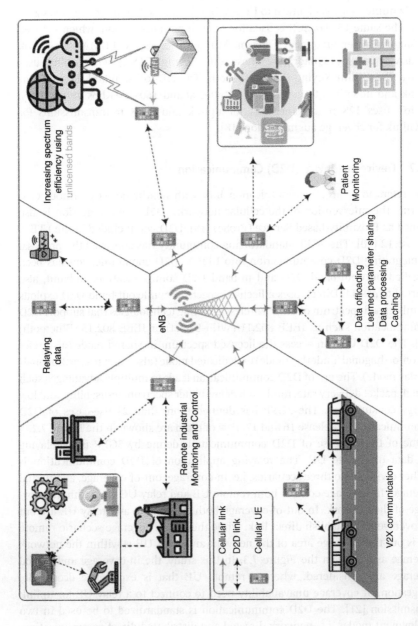

Figure 7.2 Few use cases of D2D communication.

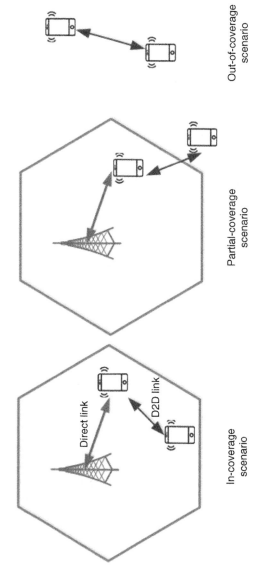

Figure 7.3 D2D communication relay scenarios.

lie in-coverage or out-of-coverage, eNB controls the discovery process and by allocating specific resources and providing the suitable relay UE to remote UE. The network discovery process involves certain exchange of synchronization, control, network authorization, discovery resource allocation, and allocating the suitable relay UE to remote UE. The number of control and synchronization messages in network directed discovery process decreases the energy efficiency and increases the latency as well. In distributed decentralized discovery process, the 3GPP has defines two remote UE discovery models, i.e. Model A and Model B [22]. Model A constitutes one-way process, where the relay UE broadcast the discovery messages within its proximity and inform remote UE of its presence and connectivity. On receiving these messages, the remote UE utilizes the information within these broadcast messages to select relay UE and establish connection over Side Link (SL). Just as uplink and downlink, the 3GPP termed the connection link between remote UE and relay as SL [23].

The Model B constitutes a two-way process, the remote UE broadcast solicitation messages to relay UE within its proximity. On receiving the solicitation message, the relay UE transmits the response to remote UE carrying information. The relay UE on receiving the response select the relay UE based on the information in the response and establish connection over SL. The direct discovery process is depicted in Figure 7.4. In our study we have considered the direct discovery process and focused on the selection process of relay UE. Model A is utilized which enable the remote UE to keep the track of relay UE.

7.2.3 Machine Learning

The Artificial Intelligence (AI) enables machines to mimic the human brain-like intelligence. The capabilities of AI include natural language processing, knowledge-based decisions, and perception. ML is the subset of AI. ML is the general technique of AI that can learn directly from structured and unstructured data provided by the information technology without any explicit programming. The ML techniques that can learn from labeled and unlabeled data sets for prediction are termed as supervised and unsupervised learning, respectively. The ML techniques enable the machines to learn themselves without any prior knowledge related to data set by interacting with the environment itself just like humans. Such ML techniques are termed as Reinforcement Learning (RL). Therefore, it classifies the ML into three categories that are supervised, unsupervised, and RL.

7.2.3.1 Reinforcement Learning

The learner (agent) in the RL kind of ML is unaware of the best course of action to take to maximize the numerical reward (to move in the direction of the main objective). However, the agent must use a hit-and-trial approach to

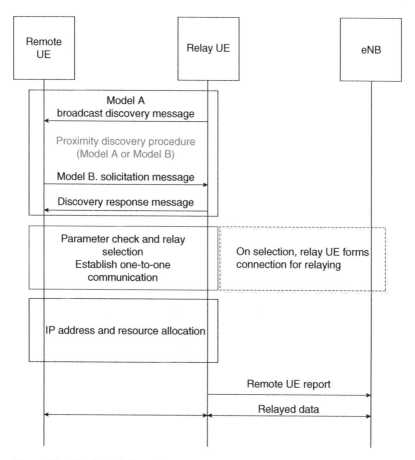

Figure 7.4 Distributed direct discovery process.

determine which action will produce the most reward. The reason for using RL is the same since it is unknown how the network environment functions in its whole. Heuristic methods, such as dynamic programming, are used to identify the best solution when the network dynamics are understood beforehand. The three learning methods used in RL are Temporal Difference (TD), Dynamic Programming (DP), and Monte Carlo (MC). To get precise results (reward) at the conclusion of the episode, policy should be modified according to MC procedures. To get the best outcomes in DP, the whole dynamics of the environment should be known. By taking action at each time step and evaluating the action in light of the reward at the end of each time step, the TD modifies the learning process. In TD approaches, the agent also lacks prior environmental knowledge [24].

In order to solve the problem using RL, the problem should be modelled as Markov Decision Process (MDP) using four tuple Bellman equation, i.e. (K, A, P, R), where K represents finite state-space, A is a finite action space of agent (a set of possible actions), P is the transitional probability matrix that determines the probability of transition from current state $k_n(t)$ to next state $k_n(t+1)$, and the R represents the reward function which determines the reward for the agent while moving from one state to another. The policy π is a sub-element of RL, which is to set the rules for an agent to select an action against the state of the environment. An RL agent at each time step t observes the state $k_n(t)$, takes an action a_t, receives the reward $r(k_n(t), a_t)$, and observes the new state $k_n(t+1)$. The goal of RL agent is to develop an optimal policy $\pi(a_t|k_n(t))$ to know the dynamics of the environment to take best action on a state for optimal solution [25].

7.2.3.2 Q-Learning

QL is type of RL that follows TD off-policy method [26]. QL is also action-value based method which in QL is known as Q-value function. The optimal policy can easily be deduced from optimal action-value function (Q-value) by selecting the maximum action-value at each state. This methodology is known as action-value based learning, as the optimal policy is derived from the action-value function. Off-policy refers to the behavior of the agent, which directly optimizes the action-value Q independently from the policy. This approach streamlines the algorithm and enables quick convergence. The policy determines and updates the action-value of the state-action pairs that are conducted in each iteration as a lookup table using the Bellman equation [24].

$$Q(k_n(t), a_t) \leftarrow Q(k_n(t), a_t) + \alpha[\triangle Q(k_n(t), a_t)]. \tag{7.1}$$

$$\triangle Q(k_n(t), a_t) = [r_k(t) + \gamma \max_a Q(k_n(t+1), a_t) - Q(k_n(t), a_t)]. \tag{7.2}$$

where is the step size, which is also known as the learning rate and the value of $\alpha \in (0; 1]$. It can be seen from Eq. (7.1) that if α is set to 0, then the agent will not learn, and when α has a high value such as 0.9, then the agent will learn quickly. $\triangle Q(k_n(t), a_t)$ is the 1-step error estimation with respect to the optimal Q-value function. It improves the action-value Q_t one step closer to the desired optimal action-value by minimizing expected value of error estimation. The γ is the discount factor to keep the reward bounded, where $0 \leq \gamma \leq 1$. The discount factor determines the present value of future rewards. When the value of γ is set to 0, the agent is more concerned about the immediate reward, that is, $r_k(t)$. As the value of γ approaches 1, the agent takes the future reward into consideration, which is the reward over the long run [27].

7.3   Related Works

Large-scale research has recently been done to improve the reliability of the 5G communication systems, which correlates to the PDR of the system by adopting D2D communication. A novel piece has been made available in [28]. It is suggested to use effective D2D communication to enhance the uploading of compliant content. In order to create a D2D communication link securely, a trust-based solution for the NB-IoT network is developed that takes into account a device's prior reputation. The suggested method seeks to weed out doubtful individuals and prevent unsuccessful broadcasts.

Petrov et al. [29] proposed NB-IoT enabled opportunistic crowdsensing based application. The automobiles serve as a relaying mechanism to take use of the D2D communication link. The suggested plan is an opportunistic paradigm in which energy-rich vehicles are outfitted with cutting-edge communication modules to help the IoT gadgets that are limited by battery life. In [30], ElGarhy and Reggiani investigated the impact of mutual interference in D2D NB-IoT UE and cellular UE (CUE). Due to the fact that CUE and NB-IoT UE send data within the same resource block, the system's spectral efficiency is increased. Elsawy et al. presented the analytical model for D2D communication [31]. For uplink cellular communication, this work presents an adaptive mode selection model with shortened channel inversion power control. Yang et al. created a method to harvest energy from BS for relay devices for D2D communication [32].

In [15] a D2D communication model based on dynamic programming has been introduced. This model employs a UE from the D2D relaying group to send data from an NB-IoT device to BS. In order to maximize reliability in terms of the predicted delivery ratio and to optimize EED, the research article formulates optimization issues. In this work, an available CUE, acting as a relay in duty cycles in an opportunistic manner, is permitted to establish a D2D link with an NB-IoT UE. The UE retransmits on the following scheduled relay node from the relaying group if the transmission is failed. After a predetermined amount of waiting time, the data packet is dropped. Such an opportunistic strategy would cause a considerable delay, a significant rise in system overhead, and a decline in system efficiency due to increased energy usage. For time-sensitive applications, such as surveillance and monitoring in smart industries, vital patient readings for heart patients in smart hospitals, and traffic management systems in smart cities, such overheads and delays are intolerable.

For relay selection, authors in [14] suggested a deterministic strategy as opposed to an opportunistic model. The suggested approach chooses the relay for D2D communication at BS, eliminating the extra latency caused by waiting for Cellular UE (CUE) to act as a relay in the opportunistic model. However, every time an NB-IoT UE has data to upload to the eNB/BS, it must emit a pilot signal in order to choose the relay in a predictable manner. For the purpose of choosing the best relay, the CUEs that are qualified and accessible for D2D communication send the pilot signal to the eNB/BS. After rating the relays in decreasing order according to channel gain and residual power, the eNB/BS chooses the best candidate to transmit the data.

7.4 System Model, Problem Formulation, and Proposed RL-ID2D

7.4.1 Network Model

In this study, a two-tier network topology with an in-band-operating NB-IoT UE and a single eNB per network cell is taken into consideration. The R UEs (including cellular and NB-IoT) are dispersed at random and are capable of direct communication with the eNB, which is positioned in the cell's center as depicted in Figure 7.5. The network model illustrates the case in which the NB-IoT UE consists of IoT devices that must communicate sensitive data. It is crucial to transmit data in a timely and trustworthy manner. This study focuses on uplink transmissions, in which an NB-IoT UE must send the data it has collected to the eNB. The maximum number of hops for data uploading is two, which means an NB-IoT UE can either upload data directly to the eNB or first transfer the data to a neighboring cellular UE (CUE). As demonstrated in Figure 7.5, the earlier example uses single-hop communication, while the latter situation uses two-hop communication through a D2D link and a cellular link. Each CUE connects with the eNB utilizing one of the C sub-channels that make up the uplink cellular spectrum. Additionally, the eNB's uplink resources are shared by the CUE and NB-IoT D2D links. To help the NB-IoT UE upload data to eNB, any eligible and available CUE in the cell can serve as a relay for D2D communication.

7.4.1.1 Channel Model
All cellular and D2D link transmissions are thought to exhibit the route loss effect according to the general power-law propagation. All cellular and D2D link transmissions are thought to exhibit the route loss effect according to the general power-law propagation. The Rayleigh channel, with a regularly distributed channel gain, is the channel model between the transmitter and receiver taken into consideration in this study. Urban environments with multipath fading of transmitted signals correlate to Rayleigh channels.

7.4.1.2 Mobility Model
The Random WayPoint model (RWP) is the mobility model used into the system model. The UEs are considered to be scattered at random around the eNB, which

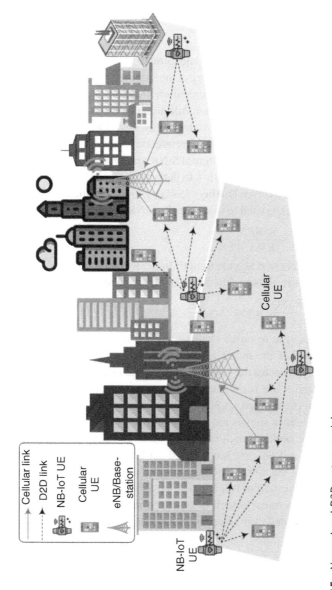

Figure 7.5 Narrowband-D2D system model.

is believed to be at the center of the cell. The UEs are all similar, dispersed randomly, and travelling at random velocities. The duration of the pause, which indicates how static the UE is at a given period, is similarly arbitrary. Additionally unpredictable is the UE's motion's direction. The CUEs are mobile, however, the NB-IoT UE is regarded as static.

7.4.1.3 Signal-to-Interference-Noise-Ratio (SINR)

Given that both D2D and cellular links use the same network resources, cross-tier interference between all of the communication lines can be avoided provided that each UE is given a separate sub-carrier. Additionally, the transmission power for D2D links is lower than that of CUE straight links, providing the distance between the D2D pair is short, which reduces interference. According to 3GPP specifications, each UE calculates its own internal SINR for the LTE-A link and reports it to the eNB during uplink transmission to determine the link quality of each UE. The Reference Signal Received Quality (RSRQ), which is defined by the Reference Signal Received Power, is used to calculate the SINR (RSRP) [33]:

$$RSRQ = N_{PRB} \times \frac{RSRP}{RSSI}. \tag{7.3}$$

where $RSSI$ is the Received Signal Strength Indicator (RSSI) and N_{PRB} is the number of physical resource blocks. The SINR is then measured as

$$SINR = \frac{12.RSRQ}{x} \quad where \quad x = \frac{RE}{RB}. \tag{7.4}$$

where RE indicates the resource element and RB indicates the resource block. The RE is composed of one sub-carrier, whereas RB is containing 12 sub-carriers. The 12 in Eq. (7.2) indicates the 12 sub-carriers of RB over which the RSSI is measured, whereas, the RSRP is measured over a single RE. In LTE-A, the Channel State Information (CSI) based on SINR is calculated by UE with these reference signals. These reference signals are present within the primary synchronization signal (PSS) and secondary synchronization signal (SSS). In case of remote-relay network, the PSS and SSS are broadcasted on sidelink synchronization signal (SLSS) and physical sidelink broadcast channel (PSBCH) with a periodicity of 40ms. The SLSS comprised of PSS and SSS, i.e. primary sidelink synchronization signal (PSSS) and secondary sidelink synchronization signal (SSSS). The PSSS and SSSS are transmitted over adjacent time slots of same subframe [21].

7.4.2 Definitions

7.4.2.1 Packet Delivery Ratio

A crucial performance parameter for assessing reliability is the PDR. The ratio of packets sent at the transmitter end to packets received at the receiver end is known as the packet delivery ratio, or PDR. The PDR is defined as follows by the phrase

$$PDR = \frac{\sum_{e=0}^{E}(E_e)}{\sum_{l=1}^{L}(L_l)}. \tag{7.5}$$

where E is the total number of packets received at the receiver and L is the total number of packets broadcast.

7.4.2.2 Potential Relay Set

The N number of CUEs that are within range of the NB-IoT UE for D2D communication and help the NB-IoT UE forward the packet to the eNB is the Potential Relay Set (*PRS*) for NB-IoT UE. The *PRS* is also denoted by the state-space $K = \{k_1, k_2, ..., k_N\}$ of the environment, where $n \in \{1, 2, ..., N\}$.

7.4.2.3 End-to-End Delivery Ratio

A performance indicator for multi-hop transmissions is the EDR. The PDR from the NB-IoT UE to the CUE relay and from the CUE relay to the BS/eNB produced it. Both ratios should be 1 to achieve 100% EDR. The following formula is used to determine the EDR:

$$EDR = \prod_{n=1}^{N}(PDR_{UE \rightarrow k_n}^{(k_n)} \cdot PDR_{k_n \rightarrow eNB}^{(k_n)}). \tag{7.6}$$

7.4.3 Problem Formulation

The problem to maximize EDR for the uplink of NB-IoT systems can be formulated as:

$$\text{maximize} \quad EDR_{UE \rightarrow CUE_{k_n} \rightarrow BS/eNB}^{l(k_n)}, \tag{7.7}$$

subject to:

$$C_1: P_{l_{(UE \rightarrow k_n)}}^{NB-IoT} \leq \zeta$$

$$C_2: \delta_{k_n} \in \{0, 1\}.$$

$$k_n \in \{k_1, k_2, ..., k_N\}, \quad l \in \{1, 2, \cdots, L\}$$

$$\delta_{k_n} = \begin{cases} 1, & SINR_{k_n \rightarrow eNB}^{(k_n)} \geq \beta \\ 0, & \text{otherwise} \end{cases}.$$

In the optimization problem Eq. (7.5), the objective function is related to the total uplink EDR over k_N relays for L packets over time steps t. The constraint $C1$ shows the required transmission power for the NB-IoT UE needed to transmit the data to the CUE relay. It also indicates that the CUE relay is bounded to be present in the feasible transmission power range ζ of the NB-IoT UE. The constraint $C2$ reflects a binary allocating indicator δ_{k_n}, that is, δ_{k_n} is 1 if the SINR of the CUE relay k_n with eNB is above the threshold β. When δ_{k_n}, CUE k_n can be included in PRS. It also prevents the use of a relay with a poor CSI to minimize the delay in the uplink transmission.

In order to solve the problem in Eq. (7.5), optimization techniques such as dynamic programming, exhaustive search, and branch and bound can solve effectively. However, the dynamics in wireless network is unknown and to determine the solution for such problems are usually NP-compete or even NP-hard [34]. To solve the problem this study proposes two-step RL based RL-ID2D as explained in Section 7.4.4.

7.4.4 Reinforcement Learning Enabled Relay Selection

The approach for dynamic relay selection proposed in this study is based on RL. The suggested method gains insight into the CUE relay's behavior by considering its availability and the accompanying EDR. Following the learning phase, the suggested RL mechanism carefully chooses the best CUE relay that offers the highest EDR. Maximum EDR, which changes with location and SINR, is the qualitative metric used to choose the relay. The best CUE relay must be chosen in order to obtain the best EDR for dependability and minimal energy consumption. In order to select the appropriate CUE relay, this part models the learning process as a QL policy.

7.4.4.1 Q-Learning Framework

The QL is made up of an agent (NB-IoT UE) that observes how the states (CUE relays) behave in the environment, a policy (parameters that specify the ideal CUE relay, or the one with the largest EDR), a reward, and the action-value function Q (accumulated reward). The policy controls how an agent (NB-IoT UE) behaves and learns at a specific time step. The suggested QL-based D2D (RL-ID2D) communication model for NB-IoT is depicted in Figure 7.6 together with its component parts. The following QL parameters are taken into account in this study:

Policy Maximizing the cumulative reward, which can be calculated as an action-value function of the Q-value using Eq. (7.1), is the primary goal of policy π in QL. It is crucial for the NB-IoT UE to choose the CUE relay that maximizes the EDR since the Q-value measures how effective a state (CUE relay) of an environment is in terms of EDR.

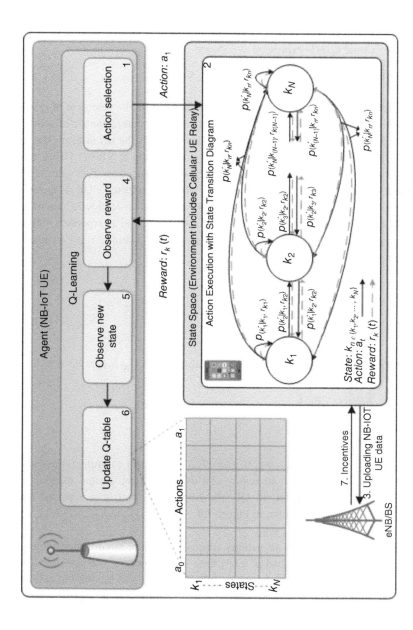

Figure 7.6 The agent-environment interaction in a Markov decision process using Q-learning, and state-transition diagram in RL-ID2D with k_N states.

State-Space The environment in this problem is the N number of states (CUE relays), that is, $k_n \in \{k_1, k_2, k_3, \ldots, k_N\}$, at discrete time step $t \in \{0, 1, 2, 3, \ldots, \}$. At each time step t, the NB-IoT UE (agent) observes the state k_n, that is, $k_n \in K$, and takes an action a_t such that $a_t \in A$ according to the policy π. Mathematically attributes of a state $k_n(t)$ is expressed as follows:

$$k_n(t) = (P^{NB-IoT}_{t_{(UE \to k_n)}} \leq \zeta, \delta_{k_n} = 1) \in K \quad \forall k_n \in PRS. \tag{7.8}$$

Action and Reward The action a_t in RL-ID2D is defined as the selection of relay UE $k_n(t)$ from PRS. The reward is the quantitative performance metric of action $a_t(k)$ on a particular state. In this study, the reward $r_k(t)$ is the reward for choosing the CUE relay k_n at time step t, which uploads the data of NB-IoT UE to the BS/eNB in a two-hop manner by exploiting D2D communication. Two values for reward are considered: 1 or 0. The reward is 1 when the selected CUE relay successfully uploads the data of NB-IoT UE to the BS/eNB, and an acknowledgment is received. Otherwise, the reward is 0. The a_t and $r_k(t)$ are related as expressed in the action-value function (Q-value Eq. (7.1)). Mathematically a_t and $r_k(t)$ is expressed as follows:

$$a_t = (k_n(t) \in K). \tag{7.9}$$

$$r_k(t) = \begin{cases} r^+ = 1, & \sum_{l=1}^{L} EDR^{l(k_n)} = 1 \\ r^- = 0, & otherwise \end{cases} . \tag{7.10}$$

Transitional Probability The transitional probability matrix $P_{kk'}$ is expressed as follows:

$$P_{kk'} = \begin{bmatrix} P_{11} & \cdots & P_{1t} \\ \vdots & \ddots & \vdots \\ P_{N1} & \cdots & P_{Nt} \end{bmatrix}$$

The rows represent states k_N, where $N = \{1, 2, \ldots, N\}$ and columns represents action a taken at time step t. In reality, the dynamics (transitional matrix) is unknown, as explained earlier. In QL, the transitional matrix is replaced by Q-value function, which is shown in Figure 7.6.

Q-Value Function The reward an activity receives from the environment determines whether it is good or bad at each time step. However, the action-value function, which is the Q-value, determines the long-term productivity of action in the given state. The cumulative reward that an agent receives at a specific state over time is specified by the Q-value of a state-action pair. It is possible for a state to have a low reward right away but a great payoff over time. Using Eq. (7.1), the Q-value in QL is computed and updated.

Convergence of QL The policy will eventually become arbitrarily near to the optimal policy since the convergence of the QL is assured. The following two restrictions determine whether the convergence occurs [24]:

- While approaching zero, the learning rates must do so gradually. Formally, this means that while their squares must converge, the sum of the learning rates must diverge.
- Each state-action pair must be visited infinitely often. This has a precise mathematical definition: each action must have a non-zero probability of being selected by the policy in every state, i.e. $\pi(a|k_n) > 0$ for all state-action pair. In practice, using an ϵ-greedy policy (where $\epsilon > 0$) ensures that this condition is satisfied.

ϵ-Greedy Policy The agent should prioritize past learnt behaviors that have produced efficient rewards in order to maximize reward; this is known as exploitation. Exploration is the process whereby the agent tries out several activities at random in an effort to choose the best course of action. Using *epsilon*-greedy is one of the greatest strategies to balance exploration with exploitation. The agent will exploit with a probability of $1 - \epsilon$ and explore with a probability of *epsilon*. Premature convergence of the system is avoided by the *epsilon*-greedy. Additionally, it increases the likelihood that new activities will be chosen. The agent chooses the action during exploitation using Eq. (7.1) in accordance with the following equation:

$$a_{greedy} = arg \max_a Q(k_n(t), a_t). \tag{7.11}$$

Incentive for CUE Relay The Smart Media Pricing (SMP) framework [35] defines the incentive for CUEs functioning as the relay. In actuality, the CUEs are egotistical and reluctance to share their resources for energy and communication. The SMP relay framework suggests that in order to combat selfishness, the relay device will charge for the energy, computation, and communication resources used during relay transmission. The source, which in the uplink or downlink scenarios will be an NB-IoT device, will pay the price as compensation for the relay. The incentives might be categorized as gratuitous service time or future-useable virtual points. The relay device is motivated to boost the resources it offers to use on the relay transmission by receiving greater incentives from the source. In this article, the incentive is given by eNB to CUEs serving as relays, however the cost of these incentives will be borne by NB-IoT UE.

7.4.5 Proposed Intelligent D2D Mechanism

For the uplink transmission, NB-IoT UE decides its association as follows. When NB-IoT UE turns on, it establishes a link with its nearest eNB. Each NB-IoT

UE sends CSI to the eNB to check channel quality. The CSI indicates SINR, as explained in Section 7.4.1.3. If SINR is over a stated threshold value, it forms a direct communication link with the eNB. Otherwise, the UE operates in D2D mode. In this paper, when NB-IoT UE is in out-of-coverage area, it requires to upload critical data to eNB and will always transmit its data in a two-hop manner with the help of CUE. To start sending the data via D2D link, this paper proposes an RL-ID2D mechanism based on the RL model. The proposed RL-ID2D mechanism is elucidated in Algorithms 7.1 and 7.2. The RL-ID2D works in two stages. The following two steps are followed by NB-IoT UE to select the best relay for D2D communication [36]:

Step 1:
The first step refers to "optPRS" (Algorithm 7.1) and provides insight into our proposed solution. The NB-IoT UE has to formulate PRS for which a solicitation signal is broadcasted within its range according to Model B, as shown in Figure 7.4. The CUE analyzes the solicitation signal and transmits the discovery response to solicitation signal. The NB-IoT UE analyzes the discovery response and analyzes by computing SINR of the communication link of CUE with eNB using RSRQ (as explained in Section 7.4.1) and compare it with SINR threshold β. The NB-IoT UE also analyzes the required transmission power of NB-IoT UE $P_t^{(NB-IoT)}$ and compare it with threshold ζ. If the CUE is eligible for D2D communication, it is including in the PRS, else it will withdraw from the selection process. The NB-IoT UE formulates a PRS with N number of CUE relays and considers PRS as the state-space k of the environment.

Algorithm 7.1: Step 1: Optimal selection of PRS (optPRS).

1: **INPUT:** (β, ζ, $P_{t_{(UE \to k_n)}}^{NB-IoT}$, *SINR*)

2:   **for** ($m = 1, \ldots, M$) **do**

3:     **if** *SINR* $> \beta$ **then**

4:       **if** $P_{t_{(UE \to k_n)}}^{NB-IoT} \leq \zeta$ **then**

5:         Push $r(t)$ UE in a new array *PRS*

6:       **else**

7:         withdraw from selection process

8:       **end if**

9:     **else**

10:       withdraw from selection process

11:     **end if**

12:   **end for**

13: **OUTPUT:** (PRS $K = \{k_1, k_2, \ldots, k_N\}$)

Algorithm 7.2: Intelligent device-to-device communication (RL-ID2D).

1: **INPUT** $(PRS, \alpha, \gamma, Q(k_n, a), \sigma)$
2: **Initialize:** $(Q(k_n, a), \alpha =$step size, $\sigma, \gamma)$
3: Select the first $k_n(t)$ randomly from *PRS*
4: **for** (*all data packets, L*) **do**
5:    $\epsilon =$ random $([0 \rightarrow 1])$
6:    **if** $\epsilon \leq \sigma$ **then**
7:       choose $k_n(t)$ randomly from the PRS for exploration
8:       **if** transmission successful **then**
9:          $r_k(t) = 1$
10:       **else**
11:          $r_k(t) = 0$
12:       **end if**
13:    **else**
14:       choose $k_n(t)$ with highest Q-value using Eq. (7.11)
15:       **if** transmission successful **then**
16:          $r_k(t) = 1$
17:       **else**
18:          $r_k(t) = 0$
19:       **end if**
20:    **end if**
21:    update $Q(k_n(t), a_t)$ using Eq. (7.1) $Q(k_n(t), a_t) + \alpha[\triangle Q(k_n(t), a_t)]$
22:    update $k_n \leftarrow k_n(t+1)$
23: **end for**

Step 2:

Algorithm 7.2 provides detailed insight into the proposed RL-ID2D. In the RL-ID2D, the NB-IoT UE inputs the information of *PRS* into the QL-based relay selection mechanism, which is explained as follows:

– The step size α such that $\alpha \in (0, 1]$, discount factor $0 \leq \gamma \leq 1$, and $\epsilon > 0$ are set.
– Initializes $Q(k_n(t), a_t)$ for all states $k_n \in K$ and action $a \in \mathscr{A}$ except for a terminal state, which is $Q(terminal) = 0$.
– Select the initial relay UE randomly as an initial state.
– For each time step t, ϵ is randomly generated between $\{0 \rightarrow 1\}$. If the value of $\epsilon \leq \sigma$, the NB-IoT UE explores to find the best relay UE by selecting the relay UE randomly from *PRS*. Otherwise, it exploits by selecting the relay UE that has the maximum Q-value from the Q-matrix using (7.11).
– The reward at each time step is observed, whether it is exploration or exploitation, and the reward matrix is updated. It should be noted that the reward is 1

for successful transmission of NB-IoT UE data to BS/eNB with good PDR; otherwise, the reward is 0.

– After the reward is recorded, RL-ID2D updates the $Q(k_n(t), a_t)$ using (7.1).
– After the learning period ends, the RL-ID2D eventually selects the best relay that maximizes the EDR with minimum overhead.

The proposed RL-ID2D algorithm works in a distributed decentralized manner at the NB-IoT UE to select the relay node for D2D communication with maximum EDR.

7.5 Performance Evaluation

In this study, we developed an LTE-A based system level simulation platform using MATLAB [37]. This section outlines the performance evaluation of the proposed RL-ID2D scheme and provides a comparison with the state-of-art techniques. The simulation scenario is explained as follows;

7.5.1 Simulation Deployment Scenario and Analysis

The simulation setup consists of R users that are independent, and randomly distributed in a single cell. Channel conditions are dynamics based on SINR value. First, we segregate the R users into NB-IoT and M number of CUE relays, and then select one of the NB-IoT users that tries to upload its data packets to the eNB using the selected CUE relay. The PRS is formed using step 1 of the Algorithm 7.1. The NB-IoT is deployed in in-band mode, in which a PRB of 180 kHz is allocated for NB-IoT UE data within cellular band. In this study, it is assumed that upon selection of CUE relay, the PRB is randomly assigned to NB-IoT UE within CUE's frame. The maximum transmit power of the NB-IoT UE is set to 14 dBm, which is standardized by 3GPP for NB-IoT to limit the interference with cellular UEs. The SINR threshold is assumed to be 13 dB. The agent in RL model learns by interacting with the state-space of environment at discrete time steps, therefore, simulation results are presented over time steps (No. of iterations). Table 7.1 shows different simulation parameters.

7.5.1.1 Analysis of Q-Learning Behavior in NB-IoT UE
Figure 7.7 illustrates the impact of considering QL for the selection of the best relay in the context of EDR. Figure 7.1 depicts the results with different values of (α, γ, σ). The increasing value of σ shows that the agent (NB-IoT UE) will explore the environment more in search of a better CUE relay. The curve shown in purple indicates that the agent learning at the rate $\alpha = 0.1$ with $\gamma = 0.3, \sigma = 0.7$

Table 7.1 Simulation parameters

Symbol	Value
No. of UEs R	50–80
Max. communication range	130 meter
SINR threshold (β)	13 dB
Path-loss exponent (μ)	3.5
Channel model	Rayleigh
Mobility model	Random Waypoint
Velocity interval	$[0.2\ 2.5]\ m/s$
Walk interval	$[1\ 2]\ s$
Pause interval	$[1\ 2]\ s$
Noise power density	-174 dBm/Hz
Max. transmission power of NB-IoT UE (ζ)	18 dB
Max. transmission power of CUE	23 dB
No. of packets (L) with size	1000 packets of 32 bytes
Exploration and exploitation constant (σ)	$0 < \sigma \leq 1$
Discount factor (γ)	$0 \leq \gamma \leq 1$
Learning rate or step-size (α)	$\alpha \in (0, 1]$
Simulation iterations	1000

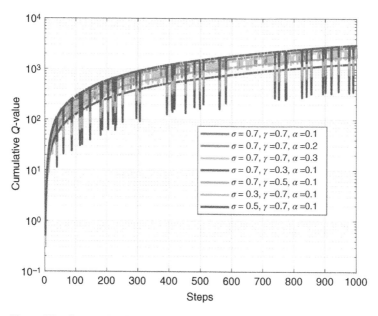

Figure 7.7 Cumulative Q-value vs. steps.

converges most quickly between 80 and 100 iterations. The downward spikes in the graph show the penalty received when the selected CUE relay fails to deliver the packet. The failure is caused due to the selection of CUE relay while exploration or exploitation, where the selected CUE UE either has the poor SINR or unavailable at the time of uplink transmission due to dynamic conditions. It is evident from the graph that as the exploration σ, the learning rate α, and discount factor γ increases, i.e. value approaches 1, the agent explores and learns more with interest in long term reward for better CUE relay selection. Therefore, the cumulative Q-value increases and the downward spike due to loss in EDR also decreases. The convergence of the graph also shows the convergence of policy toward optimal selection, that is, the CUE relay is available and provides a good EDR.

Figure 7.8 depicts the adaptive nature of RL-ID2D in dynamic environment. The EDR achieved in dynamic channel on every iteration and adaptiveness of of RL-ID2D is shown. The result is simulated on fixed 18 dB transmission power of Nb-IoT UE with $\sigma = 0.1, \alpha = 0.8, \gamma = 0.9$. The EDR at every iteration is shown by squared dot. In case of the transmission failure, the EDR drops to zero %, which can be seen as a gap circled at iteration number 190 in Figure 7.2. The failure occurs when agent (NB-IoT) selects a state (CUE relay) while exploration or exploitation, which either has the poor SINR or is unavailable at the time due

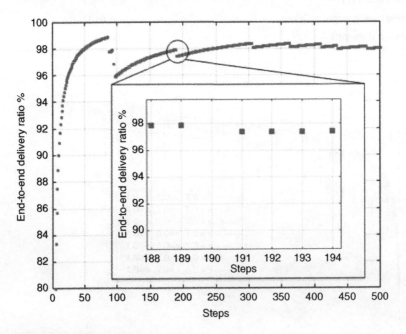

Figure 7.8 EDR vs. steps.

to dynamic conditions. A zoomed-in version is included within the Figure 7.8 for better visual of iteration number 190. RL-ID2D quickly adapts and maintain the EDR closer to 98% in the very next iteration. Similar behavior can be seen in the later iterations. The quick adaptiveness is due to the Algorithm 7.1 which makes sure to update the PRS with available and eligible CUE relays every 40 ms after receiving the reference signals broadcasted from CUEs in synchronization signals. It can be seen that as the model converges at 100th iteration, the proposed methodology also converges to select best cellular relay with optimal EDR. The EDR becomes stable more or less at 98% after system converges at 100th iteration.

Figure 7.9 shows a comparison of achieved EDR using RL-ID2D for different parameters of QL. Increasing the exploration by increasing the value of σ and increasing learning rate α, and discount factor γ allows the agent (NB-IoT UE) to learn more and to search for a better CUE relay, which is essential in a dynamic environment and yields a better-accumulated reward in the long run. Increasing the exploitation allows the agent to take the action based on past experiences, which in this case is the Q-value. The exploitation focuses on maximizing the immediate reward and allows the agent to act greedily. The uncertainty in exploration is that the action that produces a better reward is unknown. However, it is better to explore non-greedy actions if there are many time

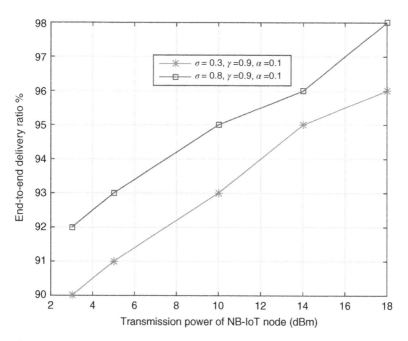

Figure 7.9 EDR vs. QL parameters.

steps ahead in which they may be subsequently exploited, which is reflected in the result, that is, increasing the exploration increases the EDR. Moreover, the result depicts that how changing QL parameter affects in achieving the near optimal EDR.

7.5.1.2 Analysis of EDR Under Various Parameters

Figure 7.10 provides a comparison of our proposed RL-ID2D with a randomly selected CUE relay for D2D communication in terms of the EDR achieved. The graph clearly shows a significant difference in the EDR achieved using both methods. Moreover, it also depicts that increasing R increases the EDR. This can be explained by the fact that increasing R increases the probability of availability of potential CUE M to be used as a D2D relay, which in turn augments EDR. In addition, the rising behavior of the curves can be understood by the fact that increasing the transmission power of the NB-IoT UE strengthens the link between the NB-IoT UE and the CUE relay, which augments the $PDR_{UE \to CUE_k}$. The results show that our proposed RL-ID2D converges to achieve an EDR of approximately 98%.

The coverage area is one of the most important parameters in communication, and directly affects the data delivery. Figure 7.11 shows the results of EDR with varying coverage areas and transmit power of NB-IoT UE $P_t = P_{t_{(UE \to k)}}^{NB-IoT}$ values.

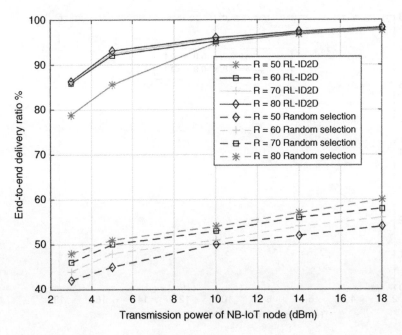

Figure 7.10 EDR vs. transmission power.

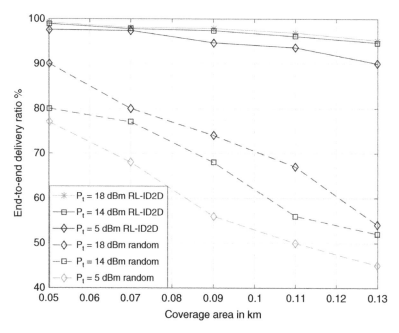

Figure 7.11 EDR vs D2D transmission range.

It also presents a comparison of random selection and RL-ID2D for a fixed number of users, $R = 50$. The result demonstrates that increasing the transmission range degrades the performance significantly when selecting the CUE relay randomly without considering the SINR and transmission power of the NB-IoT UE. While increasing the coverage area, RL-ID2D enables the NB-IoT UE to upload the data to BS/eNB with 96% EDR transmitting at 18 dBm and 90% EDR, even with a 5-dBm transmission power. This behavior is explained by the fact that the two-step RL-D2D ensures that even when $P^{NB-IoT}_{t_{(UE \to k)}}$ is low, it selects the CUE relay with optimal CQI, and QL further enhances the performance by exploration and exploitation.

7.5.1.3 Analysis of E2E Delay Under Various Parameters

Figure 7.12 shows the performance analysis of average end-to-end (E2E) delay with varying D2D communication distance over different transmission power of NB-IoT UE. The results shows the expected trend as the D2D communication distance increases the average E2E delay increases as well because increase in D2D distance promotes transmission failure and decreases EDR with increase in average E2E delay. It can also be seen from the result that as compared to low transmission power of NB-IoT UE, increase in transmission power of NB-IoT UE

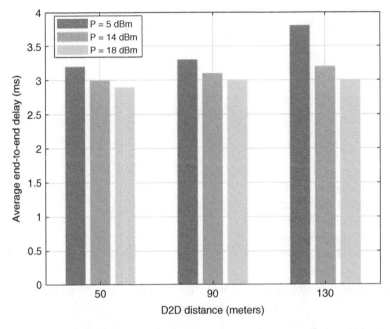

Figure 7.12 Average E2E delay vs. distance.

decreases the average E2E delay even with increase in D2D distance. This behavior can be understood by the fact that the EDR increases with high transmission power, which decreases the packet loss. The high transmission power improves the link quality between NB-IoT UE and relay UE.

Figure 7.13 depicts the performance analysis of average E2E delay at different number of cellular UE's with varying transmission power of NB-IoT UE. The decreasing trend of average E2E reflects that with increasing number of cellular UEs increases the probability of selecting the relay UE with high EDR, therefore, increase in EDR decreases packet loss rate, which decreases the average E2E delay.

7.5.1.4 Comparative Analysis of RL-ID2D with Opportunistic and Deterministic Model

Figure 7.14 provides a comparison of RL-ID2D with state-of-the-art opportunistic and deterministic schemes. The D2D communication range considered in the given result is 130 m. It can be seen that RL-ID2D outperforms other techniques, and it performs better than the opportunistic model because the opportunistic model offers the NB-IoT UE a CUE relay in an opportunistic manner. The NB-IoT UE has to seize the opportunity to upload the data in a two-hop manner; if it fails, then NB-IoT UE has to wait for the next duty cycle. It also promotes dropping the packet after a certain threshold time. Moreover, the cellular UEs

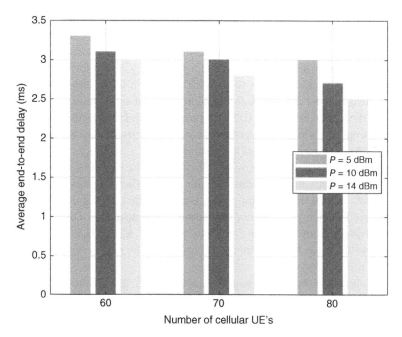

Figure 7.13 Average E2E delay vs number of cellular relays.

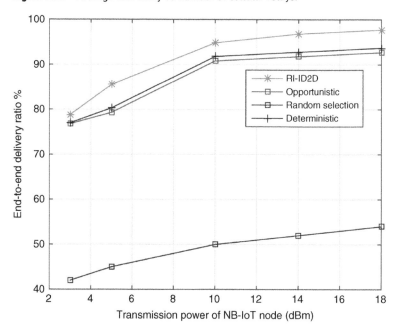

Figure 7.14 Comparison of RL-ID2D with opportunistic and deterministic model in terms of EDR.

act as a relay for NB-IoT UE in duty cycles. The deterministic model ensures the availability of the CUE relay by selecting the relay in a deterministic manner using eNB in network directed model. The eNB selects the relay after receiving the request from the NB-IoT UE to provide a CUE relay, and informs NB-IoT UE. This approach guarantees that the NB-IoT UE gets a CUE relay. However, the deterministic model augments the system overhead by increasing control signals. On the other hand, RL-ID2D improves the relay selection process by incorporating QL, which intelligently selects the optimal relay in dynamic environment. Moreover, step 1 "optPRS" guarantees the availability of the CUE relay by updating PRS periodically with CSI, as prescribed in the LTE-A standard. The opportunistic model claims that the EDR of their proposed solution approaches 98% at 8 dB transmission power. However, the D2D communication range considered in their performance evaluation is only 40 meters. On the other hand, it is evident from the Figure 7.5, our proposed RL-ID2D outperforms by achieving 98% EDR at 70 meters of D2D communication range even with 5 dB transmission power.

Figure 7.15 provides the comparison of RL-ID2D with state-of-the-art opportunistic and deterministic schemes in terms of average E2E delay. The D2D distance considered while simulating the result is 50 meters. It can be seen that

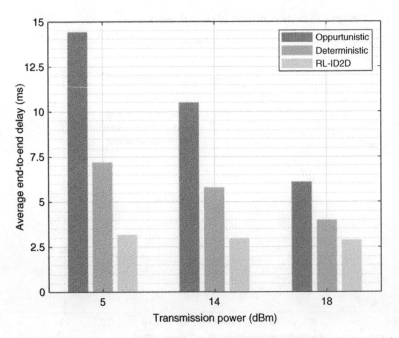

Figure 7.15 Comparison of RL-ID2D with opportunistic and deterministic model in terms of average E2E delay.

RL-ID2D outperforms other techniques. The reason is that in opportunistic model, the NB-IoT UE has to seize the opportunity to transmit the data and it promotes retransmissions, which adds delay. Moreover, the cellular UEs act as a relay in duty cycles, which further promotes packet drop and adds delay. While in deterministic model, the selection of relay UE at eNB require continuous exchange of control messages from both relay UEs and NB-IoT UE, which increases the delay. The eNB selects the relay UE with good EDR in deterministic manner at the cost of increase in E2E delay. It can be seen from the result that even at high transmission power of NB-IoT UE, the RL-ID2D outperforms opportunistic and deterministic schemes.

7.6 Conclusion

The emergence of massive MTC requires ultra-reliability in the context of data delivery with extended in-depth coverage. NB-IoT fulfills these requirements by repetitions of control and data signals. However, the fundamental solution of increased repetitions of control and data signals consumes more energy. In order to improve data delivery, a novel D2D communication link is used as a routing approach for NB-IoT, which offers the NB-IoT UE a two-hop route to reduce repetitions. An RL enabled Intelligent D2D (RL-ID2D) relay selection model is designed, which selects the cellular UE relay with the highest probability to be available with the maximum EDR and minimum E2E delay, which ultimately minimizes the EMF exposure. Simulation results depict that the proposed RL-ID2D algorithm outperforms the available state-of-the-art techniques in the literature. Cooperative communication and cooperative learning are considered to have prime role to support fully autonomous B5G and 6G network. Therefore, investigating multi-agent RL (MARL) environment, where multiple devices interact with each other and share the already learned parameter using federated learning, is an open issue, and future research should study MARL scenarios.

References

1 Zhang, Z., Xiao, Y., Ma, Z. et al. (2019). 6G wireless networks: vision, requirements, architecture, and key technologies. *IEEE Vehicular Technology Magazine* 14 (3): 28–41.

2 Statista (2015). Number of mobile phone users worldwide from 2015 to 2020 (in billions).

3 Sharma, S.K. and Wang, X. (2020). Toward massive machine type communications in ultra-dense cellular IoT networks: current issues and machine learning-assisted solutions. *IEEE Communication Surveys and Tutorials* 22 (1): 426–471.

4 Mwakwata, C.B., Malik, H., Mahtab, M. et al. (2019). Narrowband Internet of Things (NB-IoT): from physical (PHY) and media access control (MAC) layers perspectives. *Sensors* 19 (11).

5 Chen, J., Hu, K., Wang, Q. et al. (2017). Narrowband Internet of Things: implementations and applications. *IEEE Internet of Things Journal* 4 (6): 2309–2314.

6 Bay, O. (2021). 2022 Technology Summit.

7 Popli, S., Jha, R.K., and Jain, S. (2019). A survey on energy efficient narrowband Internet of Things (NBIoT): architecture, application and challenges. *IEEE Access* 7: 16739–16776.

8 Yu, C., Yu, L., Wu, Y. et al. (2017). Uplink scheduling and link adaptation for Narrowband Internet of Things systems. *IEEE Access* 5: 1724–1734.

9 Chafii, M., Bader, F., and Palicot, J. (2018). Enhancing coverage in Narrow Band-IoT using Machine Learning. *2018 IEEE Wireless Communications and Networking Conference (WCNC)*, pp. 1–6, IEEE.

10 Kar, U.N. and Sanyal, D.K. (2019). A critical review of 3GPP standardization of device-to-device communication in cellular networks. *SN Computer Science* 1.

11 Wu, H., Gao, X., Xu, S. et al. (2020). Proximate device discovery for D2D communication in LTE advanced: challenges and approaches. *IEEE Wireless Communications* 27 (4): 140–147.

12 Lianghai, J., Han, B., Liu, M., and Schotten, H.D. (2017). Applying device-to-device communication to enhance IoT services. *IEEE Communications Standards Magazine* 1 (2): 85–91.

13 Al-Tam, F., Correia, N., and Rodriguez, J. (2020). Learn to schedule (LEASCH): a deep reinforcement learning approach for radio resource scheduling in the 5G MAC layer. *IEEE Access* 8: 108088–108101.

14 Nauman, A., Ali, R., Zikria, Y.B., and Kim, S.W. (2019). An intelligent deterministic D2D communication in Narrow-band Internet of Things. *2019 15th International Wireless Communications Mobile Computing Conference (IWCMC)*, pp. 2111–2115, IEEE.

15 Li, Y., Chi, K., Chen, H. et al. (2018). Narrowband Internet of Things systems with opportunistic D2D communication. *IEEE Internet of Things Journal* 5 (3): 1474–1484.

16 Zhang, S., Liu, J., Guo, H. et al. (2020). Envisioning device-to-device communications in 6G. *IEEE Network* 34 (3): 86–91.

17 Ratasuk, R., Vejlgaard, B., Mangalvedhe, N., and Ghosh, A. (2016). NB-IoT system for M2M communication. *2016 IEEE Wireless Communications and Networking Conference (WCNC)*, pp. 1–5, IEEE.

18 Tsoukaneri, G., Condoluci, M., Mahmoodi, T. et al. (2018). Group communications in Narrowband-IoT: architecture, procedures, and evaluation. *IEEE Internet of Things Journal* 5 (3): 1539–1549.

19 Jameel, F., Hamid, Z., Jabeen, F. et al. (2018). A survey of device-to-device communications: research issues and challenges. *IEEE Communication Surveys and Tutorials* 20 (3): 2133–2168.

20 Mach, P., Becvar, Z., and Vanek, T. (2015). In-band device-to-device communication in OFDMA cellular networks: a survey and challenges. *IEEE Communication Surveys and Tutorials* 17: 1885–1922.

21 Roessler, J.S., Schlienz, J., Merkel, S., and Kottkamp, M. (2021). LTE-Advanced (3GPP Rel.12) Technology Introduction White Paper. *Rohde & Schwarz.*

22 Gamboa, S., Moreaux, A., Griffith, D., and Rouil, R. (2020). UE-to-Network Relay Discovery in ProSe-enabled LTE Networks. *2020 International Conference on Computing, Networking and Communications (ICNC)*, pp. 871–877.

23 Moreaux, A., Quintiliani, S.G., Griffith, D., and Rouil, R. (2021). UE-to-Network Relay Model B Discovery in ProSe-enabled LTE networks. 2021-04-06 04:04:00.

24 Sutton, R.S. and Barto, A.G. (2018). *Reinforcement Learning: An Introduction*, 2e. The MIT Press.

25 Nauman, A., Jamshed, M.A., Ali, R. et al. (2021). Reinforcement learning-enabled intelligent device-to-device (I-D2D) communication in narrowband Internet of Things (NB-IoT). *Computer Communications* 176: 13–22.

26 Watkins, C.J.C.H. and Dayan, P. (1992). Q-learning. *Machine Learning* 8: 279–292.

27 Melo, F.S. and Ribeiro, M.I. (2007). Convergence of Q-learning with linear function approximation. *2007 European Control Conference (ECC)*, pp. 2671–2678.

28 Militano, L., Orsino, A., Araniti, G. et al. (2016). Trusted D2D-based data uploading in in-band Narrowband-IoT with social awareness. *2016 IEEE 27th Annual International Symposium on Personal, Indoor, and Mobile Radio Communications (PIMRC)*, pp. 1–6, IEEE.

29 Petrov, V., Samuylov, A., Begishev, V. et al. (2018). Vehicle-based relay assistance for opportunistic crowdsensing over narrowband IoT (NB-IoT). *IEEE Internet of Things Journal* 5 (5): 3710–3723.

30 ElGarhy, O. and Reggiani, L. (2018). Increasing efficiency of resource allocation for D2D communication in NB-IoT context. *Procedia Computer Science* 130: 1084–1089.

31 ElSawy, H., Hossain, E., and Alouini, M.-S. (2014). Analytical modeling of mode selection and power control for underlay D2D communication in cellular networks. *IEEE Transactions on Communications* 62 (11): 4147–4161.

32 Yang, H.H., Lee, J., and Quek, T.Q.S. (2016). Heterogeneous cellular network with energy harvesting-based D2D communication. *IEEE Transactions on Wireless Communications* 15 (2): 1406–1419.

33 Parikh, J. and Basu, A. (2016). Effect of mobility on SINR in long term evolution systems. *ICTACT Journal on Communication Technology* 7: 1239–1244.

34 Huang, C.-Y.R., Lai, C.-Y., and Cheng, K.-T.T. (2009). Fundamentals of algorithms. In: *Electronic Design Automation*, Chapter 4 (ed. L.-T. Wang, Y.-W. Chang, and K.-T. T. Cheng), 173–234. Boston, MA: Morgan Kaufmann.

35 Wang, W. and Wang, Q. (2018). Price the QoE, not the data: SMP-economic resource allocation in wireless multimedia Internet of Things. *IEEE Communications Magazine* 56 (9): 74–79.

36 Nauman, A., Jamshed, M.A., Qadri, Y.A. et al. (2021). Reliability optimization in narrowband device-to-device communication for 5G and beyond-5G networks. *IEEE Access* 9: 157584–157596.

37 MATLAB (2020). *Version (R2020b)*. The MathWorks Inc.

8

Unsupervised Learning Based Resource Allocation for Low EMF NOMA Systems

Muhammad Ali Jamshed[1], Fabien Héliot[2], and Tim W.C. Brown[2]

[1] *James Watt School of Engineering, University of Glasgow, Glasgow, UK*
[2] *Institute of Communication Systems (ICS), Home of 5G and 6G Innovation Centre, University of Surrey, Guildford, UK*

8.1 Introduction

Advances in the wireless communication sector have transformed how societies originate, distribute, receive, and interpret information. In the 5G era, it is expected that capacity would have to increase 1,000 times to support the ever-growing number of wireless users and the Internet of things (IoT) devices (augmented reality, wearable sensors, etc.) [1]. This capacity increase will be made feasible by increasing the number of access points (APs). In other words, the number and variety of EM field (EMF) exposure sources in the environment will surge in the 5G era. Despite the lack of strong evidence regarding the short-term health effects of EMF exposure, the international agency for research on cancer (IARC) and the World Health Organization (WHO) have classified EMF radiation from wireless devices as possibly carcinogenic to humans (group B) [2]. Meanwhile, the rise in EMF radiations in the 5G era may only enhance the potential long-term health concerns associated with EMF [3].

The user terminal (UT) antennas must conform with EMF rules in order to reduce the potential health effects of EMF exposure. For example, the Federal Communications Commission (FCC) in the United States has local specific absorption rate (SAR) rules, stating that the maximum SAR value for 1 g of human tissue should not exceed 1.6 W/kg, and similarly, for 10 g of tissue, it should be less than 2.0 W/kg [4]. Meanwhile, the international commission on non-ionizing radiation protection (ICNIRP) has specified the limits for both passive (power density (PD)) and active (SAR) EMF exposure [5]. According to recent ICNIRP recommendations, the whole-body average SAR (i.e. global SAR) should not be greater than 0.4 W/kg. However, according to FCC consumer guidelines,

Low Electromagnetic Field Exposure Wireless Devices: Fundamentals and Recent Advances, First Edition.
Edited by Masood Ur Rehman and Muhammad Ali Jamshed.

"FCC approval does not indicate the amount of EMF exposure consumers experience during normal use of the device" [6].

Previous generations of mobile communications (2G–4G) depended on orthogonal multiple access (OMA)methods, such as orthogonal frequency division multiple access (OFDMA), time-division multiple access (TDMA), and so on, to achieve user multiplexing. The non-orthogonal multiple access (NOMA), on the other hand, has emerged as a potential multiplexing strategy for the fifth generation of cellular communication technology. There are two forms of NOMA: power domain-NOMA (PD-NOMA) [7–9] and code-domain NOMA (CD-NOMA) (e.g. sparse code multiple access (SCMA) [10, 11], and we will focus on the former in this chapter. The previous generations (2G–4G) of mobile communications relied on OMA schemes, i.e. OFDMA, TDMA, etc., to perform user multiplexing. However, the NOMA has emerged as a promising multiplexing approach for the fifth generation of cellular communication technology. There are two main types of NOMA, which are known as PD-NOMA [7–9] and code-domain NOMA (CD-NOMA) (e.g. sparse code multiple access (SCMA) [10, 11], and in this chapter we focus on the former. The PD-NOMA takes use of the differences in channel gain between users to place them on the same subcarrier with varied transmission strengths, allowing the system to schedule more users at the same time [7]. In comparison with OMA, NOMA can manage a greater number of users, boosting the system's spectral efficiency. Furthermore, the features of traditional PD-NOMA provide some attractive options to minimize EMF exposure while maintaining quality of service (QoS).

8.1.1 Existing Work

According to our latest literature survey on EMF exposure risk assessment and evaluation process in [12], the majority of existing work on EMF reduction in the uplink of cellular system boils down to SAR reduction. In [13], for example, the SAR is efficiently decreased by inserting a metamaterial between the UT and the human head. The studies aimed at reducing SAR only address the worst-case scenarios and do not account for the influence of EMF exposure over time. In [14], an EMF-aware scheduling approach for the uplink of OFDMA-based systems has been suggested, which uses the definition of the dosage metric (measurement of SAR over time) to limit user EMF exposure in a single cell system while ensuring QoS. A similar strategy is used in [15] to decrease EMF exposure in an OFDMA-based multi-cell environment. Whereas in [16], a cell-selection approach is employed to limit the uplink EMF exposure of users in a single cell scenario while meeting all of these users' downlink throughput needs.

In terms of PD-NOMA, most previous papers have concentrated on the downlink situation, with just a few considering the uplink of PD-NOMA [17, 18]. For example, in [17], throughput is increased by carefully altering the power levels of each user who is assigned to the same subcarrier, compared with the old OMA system. Similarly, the spectral efficiency of an PD-NOMA-based system is increased by employing an effective strategy to subcarrier allocation while also conducting optimal power assignment [18]. The performance of PD-NOMA systems is heavily dependent on optimal user grouping/paring on the same subcarrier, in addition to effective power allocation algorithms. In [19], a dynamic user grouping method based on the underlying Rayleigh fading channel coefficients is presented to improve detection accuracy at the receiver end by minimizing interference amongst users that share the same subcarrier. The authors employ an adaptive user pairing approach in [8] to detect a user far from the access point (AP) and then pick a user with a better channel condition to group them on the same subcarrier. Recently, machine learning (ML)-based clustering algorithms have been employed to intelligently execute user grouping for NOMA systems by determining an acceptable number of clusters [9]. The existing research on clustering algorithms classifies them as hierarchical or partitional techniques [20]. The hierarchical clustering technique is made up of layered clusters arranged as trees, whereas the partitional algorithms cluster the data in disjoint subsets [21].

8.1.2 Motivation and Contributions

In contrast to OMA, users with different power levels can be multiplexed on the same subcarrier utilizing the PD-NOMA technique. It is strongly advised that users have a low channel correlation to prevent interference between users using the same subcarrier. As a result, selecting appropriate user clustering strategies is important to realizing the full potential of a PD-NOMA system. The combinatorial nature of user grouping/clustering drives the need for ML techniques, particularly in big systems. In comparison with suboptimal methods, such as matching theory [22], the learning aspects of the ML algorithms give an additive benefit in clustering users into distinct groups based on their related channel parameters. The clustering techniques used are mostly determined by the distribution of the data set in the feature space. The unsupervised machine learning technique may be used to create a clustering learner by intelligently splitting the data set in the feature space using a measurement function.

In this chapter, we present a novel EMF-aware scheduling strategy that uses PD-NOMA and unsupervised ML to reduce cellular system users' uplink exposure even more than other current scheduling schemes while meeting QoS criteria. We employ ML to provide an effective user grouping approach that aids in successful decoding at the receiver end. We initially clustered the users using the K-means [9] clustering technique, and then we effectively grouped and distributed the users to the subcarriers to decrease their exposure. The K-means method is based on a partitional clustering strategy that divides the data into a number of clusters that the user specifies. To the best of our knowledge, no study in the literature has yet utilized the unsupervised ML idea to limit the EMF exposure of cellular systems based on the PD-NOMA technology. Our primary contributions to this chapter are as follows:

1. In contrast to [14], we design a multiuser EMF-aware scheduling based on PD-NOMA technology rather than the classic OFDMA technology used in [14], which makes the design more difficult, e.g. user grouping, but provides additional benefits in terms of exposure reduction. Jamshed et al. [23] contains a preliminary version of this work. This work, in comparison with [23], provides more robust and efficient techniques for categorizing users (based on ML) and distributing these users to subcarriers. It also integrates the effect of successive interference cancellation (SIC) in the allocation process by imposing a minimal interference restriction on users using the same subcarrier, making our solution more suited for practical settings.

2. We utilized the K-means technique to cluster the users into distinct categories, similar to [9]. In contrast to [9], we employed the elbow approach in conjunction with the F-test method for the first time to discover the best number of clusters for a specific channel condition. Furthermore, our suggested user clustering method outperforms [9] in terms of control over the number of users per subcarrier and bandwidth consumption. Furthermore, this study gives a mathematical concept for determining the F-test value for a given number of clusters produced using K-means while relying on each user's associated channel attributes. Finally, we allocate power in an EMF-aware way.

3. The suggested strategy's performance has been validated using Monte Carlo simulations. We compared our proposed method with a comparable OFDMA-based EMF-aware multiuser scheduling strategy, a spectrum efficient PD-NOMA scheme, and the greedy subcarrier allocation technique, and found that our novel EMF-aware PD-NOMA scheme successfully reduces EMF exposure. In comparison with [14], our unique strategy decreases EMF exposure by at least 28%, and by at least 2 and 3 orders of magnitude when compared with [18] and the greedy approach.

4. The SAR values derived using the simulation scenarios built in CST STUDIO SUITE 2018 based on IEEE guidelines further verified our results. Furthermore, we created an planar inverted-F (PIFA) with a single transmit antenna to replicate the characteristics of a mobile device. This sort of study also shows the superiority of incorporating PIFAs into mobile devices.

8.1.3 Structure of the Chapter

The rest of the chapter is structured as follows: Section 8.2 provides a full explanation of the system model, as well as the scenario of interest and mathematical framework that were examined when designing our EMF-aware PD-NOMA proposed scheme. Section 8.3 discusses the unique planned ML-based user categorization and resource distribution method. Meanwhile, Section 8.4 discusses our EMF-aware power allocation technique. Section 8.5 contains the simulation findings as well as the pertinent remarks. Finally, Section 8.6 brings the chapter to a conclusion. The list of symbols with their meanings, used throughout the chapter are listed in Table 8.1.

Table 8.1 Notations with definitions used throughout the chapter.

Symbol	Meaning
R	Cellular radius
β	Conductivity
α	Subcarrier index
σ^2	Noise power density per subcarrier
M_d	Mass density
K	Total number of users
T	Total number of time slots
S	Total subcarrier allocated over T
g_n	Gain of user n
W	Total bandwidth
E_f	Electric field intensity
D_k	Path loss
$P_k^{\max}$	Maximum transmit power
$\bar{p}_k$	Signaling power of kth user
τ	Duration of time slot t
a	CQI bits
ζ	Reference threshold
M	Number of clusters
z_n	Gaussian noise
P_o	Received signal power
SAR_k^a	k^{th} user SAR for cheek position
SAR_k^b	k^{th} user SAR for tilt position
Br	Required number of bits
$K_{n,t}$	Maximum number of users on α

Source: Jamshed et al. [24]/IEEE.

8.2 EMF-Aware PD-NOMA Framework

8.2.1 System Model

As demonstrated in Figure 8.1, we assume a single-cell communication system in which K users rely on single antenna UTs to interact with an base station (BS) using the PD-NOMA multiplexing approach. The system bandwidth, W, is divided into N subcarriers, and because we are considering an NOMA scenario, several users might be allocated to the same subcarrier. If $K_{n,t}$ users are allocated to the nth subcarrier during time slot t, the associated received signal at the BS can be represented as follows:

$$y_{n,t} = \sum_{k=1}^{K_{n,t}} \sqrt{p_{k,n,t} g_{k,n,t}} x_{k,n,t} + z_{n,t}, \tag{8.1}$$

where $x_{k,n,t}$ is the information signal of kth UT on subcarrier n at time slot t, $p_{k,n,t}$ is the transmit power of the kth UT on subcarrier n at time slot t, $g_{k,n,t}$ is the channel gain between the kth UT and the BS on subcarrier n at time slot t, and $z_{n,t}$ models the additive white Gaussian noise (AWGN), having a zero mean and a variance of σ^2, on subcarrier n at time slot t. Interference arises because data from various users might be multiplexed on the same subcarrier in NOMA. In the general situation, the total interference experienced by the kth user of subcarrier n can be stated as

$$\bar{I}_{k,n,t} = \sum_{l=1, l \neq k}^{K_{n,t}} p_{l,n,t} g_{l,n,t}. \tag{8.2}$$

The SIC procedure is often used at the receiver (i.e. BS in our instance) to decode the data of the several users that are multiplexed together in PD-NOMA. It is generally known from multiple-input multiple-output (MIMO) detection, for which SIC was initially created [25], that the order of the detection affects the process performance. In SIC-based MIMO detection, data supplied over the best channel gain is traditionally processed first, while data sent over the poorest channel gain is decoded last. In the uplink of PD-NOMA, this rule translates as the user having the best received power at the BS (i.e. $p_{\pi(1),n,t} g_{\pi(1),n,t}$) being decoded first, while the one with the worst power at the BS (i.e. $p_{\pi(K_{n,t}),n,t} g_{\pi(K_{n,t}),n,t}$) being decoded last, where π is an index permutation vector based on the SIC decoding order, such that $\pi(1) = \text{argmax}_k p_{k,n,t} g_{k,n,t}$, $\pi(K_{n,t}) = \text{argmin}_k p_{k,n,t} g_{k,n,t}$, and $p_{\pi(1),n,t} g_{\pi(1),n,t} \geq p_{\pi(2),n,t} g_{\pi(2),n,t} \geq \cdots \geq p_{\pi(K_{n,t}),n,t} g_{\pi(K_{n,t}),n,t}$ on any given subcarrier n of time slot t. As a result, power allocation at the UT will affect the performance of the SIC process, i.e. its capacity to dissociate between users. The multiplexing method of PD-NOMA is predicated on the assumption that SIC should be able to properly decode the multiplexed signal [26]. In order to take this into account in

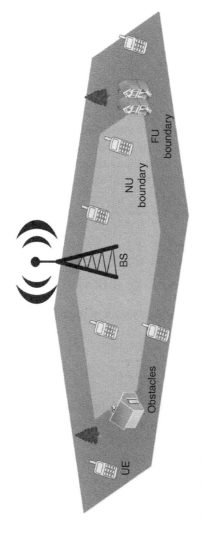

Figure 8.1 An illustration of a single cell system with multiple users. Source: Jamshed et al. [24]/with permission of IEEE.

our model, we consider that this can be achieved for any user k on subcarrier n of time slot t if it received a signal to residual interference ratio is greater than or equal to a reference threshold ζ, i.e.:

$$p_{k,n,t}g_{k,n,t}/I_{k,n,t} \geq \zeta, \tag{8.3}$$

$\zeta \geq 1$, where

$$I_{k,n,t} = \sum_{l=\pi^{-1}(k)+1}^{K_{n,t}} p_{\pi(l),n,t}g_{\pi(l),n,t}. \tag{8.4}$$

In (8.4), $\pi^{-1}(k)$ returns the index of k in the permutation vector π, e.g. for instance if $\pi(1) = 5$, then $\pi^{-1}(5) = 1$, i.e. the index of 5 in π is 1. Consequently, when the SIC process is always successful, or in other words, when the constraint in (8.3) is always met, then $I_{\pi(K_{n,t})} = 0, \forall n, t$. The total amount of bits sent by user k on subcarrier n throughout a time slot t of period τ may be estimated using the following formula based on Equations (8.1), (8.4) and Shannon theory:

$$b_{k,n,t(\alpha,p)} = w\tau\, \alpha_{k,n,t}\log_2\left(1 + \frac{p_{k,n,t}g_{k,n,t}}{\sigma^2 + I_{k,n,t}}\right), \tag{8.5}$$

where $w = W/N$ is the bandwidth of each subcarrier, $\alpha_{k,n,t}$ is a subcarrier allocation index, i.e. $\alpha_{k,n,t} = 1$ if user k is allocated to subcarrier n at time slot t or 0 otherwise, $p = [p_{1,1,1}; \dots; p_{K,N,1}; p_{1,1,2}; \dots; p_{K,N,T}] \geq 0$, and $\alpha = [\alpha_{1,1,1}; \dots; \alpha_{k,N,1}; \alpha_{1,1,2}; \dots; \alpha_{K,N,T}]$.

As far as the individual exposure of user k in the uplink is concerned, it can be expressed, based on [27], as

$$E_k(\alpha, p) = \frac{SAR_k}{TP^{\text{ref}}}\tau\left(\bar{p}_k(T) + \sum_{t=1}^{T}\sum_{n=1}^{N}\alpha_{k,n,t}p_{k,n,t}\right); \tag{8.6}$$

here, T is the number of time slots, P^{ref} is defined at the reference incident power at which the SAR of the kth UT has been calculated. The SAR metric is generally used to measure the EMF exposure in the near-field of an antenna operating below 10 GHz. It can be categorized as average or organ-specific SAR, e.g. head, hand, etc. [12], and it is usually expressed as

$$SAR = \frac{\beta \times E_f^2}{M_d} \quad (W/kg), \tag{8.7}$$

where β, M_d accounts for the conductivity and the mass density of the exposed object/tissues, respectively, and the E_f indicates the electric field. Note that the FCC enforces a limit of 1.6 W/kg for one gram of body mass for the SAR of UTs [28]. Meanwhile, $\bar{p}_k(T)$ in (8.6) represents the signaling power used by the UT of user k for setting up its data transmission over T time slots; it can mathematically be expressed, based on [29], as

$$\bar{p}_k(T) = \min\left(P_k^{\max}, P_o D_k \ \max\left(aT/4, 1\right)\right), \qquad (8.8)$$

where $P_k^{\max}$ is the kth UE maximum transmit power, D_k denotes the path loss experienced by user k, P_o denotes the received signal power threshold at the BS, and a denotes the number of channel quality index (CQI) bits used to measure the channel quality.

8.2.2 Problem Formulation

With the primary goal of designing an EMF-aware scheduling method for lowering exposure while considering QoS and PD-NOMA with SIC, we first describe its mathematical framework in the form of an optimization problem, as follows:

$$\underset{p,\alpha}{\text{minimize}} \quad E(\boldsymbol{\alpha}, \boldsymbol{p}) = \sum_{i=1}^{K} E_k(\boldsymbol{\alpha}, \boldsymbol{p}), \qquad (8.9)$$

subject to:

$$C_1 : \sum_{t=1}^{T}\sum_{n=1}^{N} b_{k,n,t}(\boldsymbol{\alpha}, \boldsymbol{p}) = Br_k, \forall k \qquad (8.10a)$$

$$C_2 : \sum_{n=1}^{N} \alpha_{k,n,t} p_{k,n,t} \leq P_k^{\max} \quad \forall k, \forall t, \qquad (8.10b)$$

$$C_3 : \alpha_{k,n,t}(p_{k,n,t} g_{k,n,t}/I_{k,n,t}) \geq \zeta \quad \forall k, \forall n, \forall t \qquad (8.10c)$$

$$C_4 : \sum_{k=1}^{K} \alpha_{k,n,t} \leq K_{n,t} \quad \forall n, \forall t. \qquad (8.10d)$$

In (8.9), the objective function $E(\boldsymbol{\alpha}, \boldsymbol{p})$ reflects the total uplink exposure experienced by the system's K users across T time slots. In the meanwhile, the constraints in (8.10) are as follows:

1. The constraint C_1 is a QoS constraint that assures that each user sends the needed number of bits, i.e. Br_k, in the appropriate order (8.10a).
2. The constraint C_2 is a transmit power constraint; it indicates that the transmit power of each UT is restricted by a maximum value, i.e. $P_k^{\max}$ in (8.10b), for each time slot.
3. The constraint C_3 is connected to the SIC procedure at the BS; if (8.10c) is fulfilled, then SIC can remove some interference from user k's received signal.
4. The constraint C_4 is related to NOMA; it reflects that the number of users that can be grouped on subcarrier n at any time slot t is at most $K_{n,t}$, when using NOMA.

It should be observed that our optimization problem, as specified in (8.9) and (8.10), is non-convex due to the binary character of $\alpha_{k,n,t}$ [30] and the non-affine equality constraint C_1. To properly solve this issue, we must first overcome its binary character by employing a typical relaxation process. Indeed, like in [31], we investigate a sequential subcarrier and power assignment, i.e. first finding α for a fixed p and then finding p for a fixed α. A variable adjustment is also required in (8.9) and (8.10) to overcome the non-affinity of C_1 and make them convex when α is fixed. These two points are discussed in further depth in the following sections. To begin, in Section 8.3, we describe our novel ML-based grouping and subcarrier allocation method, which intelligently provides complete control over the number of users grouped on the same subcarrier and reduces the complexity of SIC at the receiver end by grouping users on a subcarrier n with the least amount of interference. Second, we describe our power allocation technique in Section 8.4, which is coupled with our ML-based subcarrier allocation. It adjusts each user's transmit power level to guarantee that their QoS is met. It should be emphasized that, as in [14], we assume that the BS can forecast the channel state information (CSI) of the K users across T time slots in advance using uplink pilot signals.

8.3 Machine Learning Based User Grouping/Subcarrier Allocation

The PD-NOMA strategy helps communication systems to boost their spectral efficiency [7] by efficiently demultiplexing multiple users sharing the same subcarrier based on their received power levels. Its efficacy is mainly dependent on the receiver's capacity to distinguish between different levels of user received powers, which is dependent on the selection/grouping of the users that are multiplexed together. In the uplink of a cellular system, for example, the location of the users (near user (NU) and far user (FU) in Figure 8.1) can be used to group the users (NU and FU are expected to have relatively different received power levels at the BS owing to their considerable variation in path-loss).

When it comes to user grouping, using ML algorithms decreases complexity and enhances the likelihood of convergence to an ideal solution when compared with non-ML approaches [32]. For example, K-means clustering, one of the most basic and widely used unsupervised ML algorithms, has proven to be an efficient way for categorizing users in [9]. The K-means method divides data into various groups/clusters, with the required number of clusters set before executing the algorithm. As a result, determining the optimal number of clusters for attaining a certain goal is a crucial aspect of the clustering process [33]. A basic rule of thumb for choosing this number is to keep it modest in relation to the size of the data set to be clustered (in our example, $g_{k,n,t}$ $\forall k, n, t$). The authors of [9] employed a simplified version of the elbow approach [34] to establish their desired number of clusters in the context of PD-NOMA. This elbow technique variant use the K-means approach to compute the average internal cluster distance for a given range of desired cluster numbers [9].

Similarly to [9], we utilize the K-means approach to cluster the user, but we use the F-test method in conjunction with the elbow method to change the amount of clusters necessary each subcarrier and time slot. The F-test determines the ratio of the variability between the different clusters (i.e. produced by using K-means) to the overall variability, which is the sum of the variability between the different clusters and the variability within the clusters, for a given number of clusters M, as follows:

$$
F_{n,t}(M) = \frac{\sum_{m=1}^{M} K_m (\bar{g}_{m,n,t} - \bar{g}_{n,t})^2 / (M-1)}{\left[\sum_{m=1}^{M} \sum_{j=1}^{K_m} (g_{m(j),n,t} - \bar{g}_{m,n,t})^2 / (K-M) \right.} \\
\left. + \sum_{m=1}^{M} K_m (\bar{g}_{m,n,t} - \bar{g}_{n,t})^2 / (M-1) \right],
\tag{8.11}
$$

for each subcarrier of each time slot. In (8.11), K_m is the number of users in the mth cluster and $\bar{g}_{n,t} = \sum_{k=1}^{K} g_{k,n,t}/K$ is the mean of $g_{k,n,t}$ over the K users, $\forall n, t$. Moreover, $\bar{g}_{m,n,t} = \sum_{j=1}^{K_m} g_{m(j),n,t}/K_m$ is the mean of $g_{k,n,t}$ over the K_m users belonging to the mth cluster, $\forall n, t$, where $g_{m(j),n,t}$ represents the channel gain of the jth user belonging to the mth cluster, $\forall n, t$. Figure 8.2 depicts F-test results as a function of the number of clusters M when running the K-means algorithm [9] for each of these M values (i.e. $M = 1$ to 14) for $K = 100$ over one subcarrier/time slot (i.e. $N = T = 1$) and assuming that $g_{k,1,1}$ follows a Rayleigh distribution (by using the path The findings demonstrate that the F-test increases as M grows, indicating that the similarities between channel gains within each cluster and the differences between channel gains within clusters both increase with M. It should also be noted that, as is common in clustering research, the curve creates an "elbow" shape from $M = 2$ to 4. As a result, an appropriate value of M may be discovered inside this period using the elbow approach. In Figure 8.2, for example, a cluster number of $M = 3$ with a F value of 99% appears to be a suitable figure for M. It should be noted that requirements and constraints (for example, accommodating more users during peak hours) can also be considered (in conjunction with the elbow method) to determine an appropriate value of M; that is, first use the elbow method to find a suitable range of M values, and then select an appropriate value of M within this range based on specific requirements/constraints.

Once a sufficient M value for a specific subcarrier n at time slot t has been determined, a user from each cluster is scheduled on this subcarrier/time slot, as

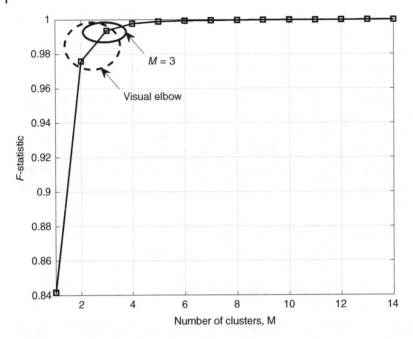

Figure 8.2 *F*-test statistic in (8.11) as a function of *M* for $K = 100$, $N = 1$, and $T = 1$. Source: Jamshed et al. [24]/with permission of IEEE.

specified in Algorithm 1. As a result, the number of users clustered on this subcarrier/time slot is $K_{n,t} = M$. The normalized channel gain is $G_{k,n,t}$ in Algorithm 8.1, which is utilized to accomplish subcarrier allocation. We employ $G_{k,n,t}$ in the subcarrier allocation instead of $g_{k,n,t}$ to be fair to all users and to limit EMF exposure by avoiding the worst subcarrier allocation to the user during a given time slot t. Furthermore, Sk represents the number of assigned subcarriers per user.

8.4 Power Assignment

Following the user grouping and subcarrier allocation processes described in Section 8.3, an optimal amount of transmit power must be assigned to each user in order to limit uplink EMF-exposure. Section 8.2.2 shows that even when α is known (from the subcarrier allocation procedure), the issue in (8.9) and (8.10) is non-convex owing to constraint non-linearity (8.10a). Using the following variable change (based on a basic transformation of (8.5)):

$$p_{k,n,t} = \frac{(2^{r_{k,n,t}} - 1)(\sigma^2 + I_{k,n,t})}{g_{k,n,t}}, \tag{8.12}$$

Algorithm 8.1: Machine Learning (ML) based user grouping and sub-carrier assignment. Source: Jamshed et al. [24]/with permission of IEEE.

1: **INPUT** $(K, T, N, g_{k,n,t})$
2: Set $\alpha = 0$;
3: Set $G_{k,n,t} = g_{k,n,t}/\bar{g}_k \; \forall k, n, t, \bar{g}_k = \sum_{t=1}^{T} \sum_{n=1}^{N} g_{k,n,t}/K/T$;
4: **Step 1:** User grouping and setting M
5: **for** $t = 1: T$ **do**
6:    **for** $n = 1: N$ **do**
7:       **for** $M = 2: K - 1$ **do**
8:          Use K-means [9] to cluster $\mathbf{g}_{n,t} = [g_{1,n,t}, \dots, g_{K,n,t}]$ into M clusters and save the clustering pattern for each value of M;
9:          Calculate F-test in (8.11);
10:       **end for**
11:       Select M as per requirement based on F-test;
12:       Set $K_{n,t} = M$;
13:    **end for**
14: **end for**
15: Set $S_k = \lfloor \sum_{n=1}^{N} \sum_{t=1}^{T} K_{n,t}/K \rfloor, \forall k$.
16: **Step 2:** Subcarrier allocation
17: **for** $t = 1: T$ **do**
18:    **for** $n = 1: N$ **do**
19:       Use the saved clustering pattern corresponding to the selected value of $K_{n,t}$ to cluster $\mathbf{G}_{n,t} = [G_{1,n,t}, \dots, G_{K,n,t}]$ into $K_{n,t}$ clusters;
20:       Pick the user having the maximum value of $G_{.,n,t}$ within each clusters and set the corresponding value of $\alpha_{.,n,t}$ to 1 if current allocated number of subcarrier is below S_k;
21:    **end for**
22: **end for**
23: **OUTPUT** α, S

and by substituting $p_{k,n,t}$ by (8.12) in (8.9) and (8.10), this optimization problem can be re-expressed as

$$\min_{r} \frac{\tau}{T} \sum_{k=1}^{K} \left(\bar{P}_k(T) + \sum_{t=1}^{T} \sum_{n=1}^{N} \frac{\alpha_{k,n,t}(2^{r_{k,n,t}} - 1)(\sigma^2 + I_{k,n,t})}{g_{k,n,t}} \right),$$

subject to:

$$C_1 : w\tau \sum_{t=1}^{T} \sum_{n=1}^{N} \alpha_{k,n,t} r_{k,n,t} = Br_k,$$

$$C_2 : \sum_{n=1}^{N} \frac{\alpha_{k,n,t}(2^{r_{k,n,t}} - 1)(\sigma^2 + I_{k,n,t})}{g_{k,n,t}} \leq P_k^{\max},$$

$$C_3 : \alpha_{k,n,t}((2^{r_{k,n,t}} - 1)(\sigma^2 + I_{k,n,t})/I_{k,n,t}) \geq \zeta, \tag{8.13}$$

where $r_{k,n,t} = b_{k,n,t}/(w\tau)$ represents the spectral efficiency of user k on subcarrier n during time slot t, and $\mathbf{r} = [r_{1,1,1}; \dots; r_{K,N,1}; r_{1,1,2}; \dots; r_{K,N,T}]$. Contrary to (8.10), the constraint C_1 in (8.13) is an affine function of $\mathbf{r}$. Moreover, the objective function and the constrained C_2 and C_3 are all convex functions of $\mathbf{r}$ in (8.13), when assuming that $I_{k,n,t}$ is fixed. Thus, (8.13) is a convex optimization problem and its Lagrangian can be defined as

$$\mathcal{L}(r_{k,t}, \lambda_k, \mu_{k,t}, \delta_{k,n,t})$$

$$= \frac{\tau}{T} \sum_{k=1}^{K} \left(\bar{p}_k + \sum_{t=1}^{T} \sum_{n=1}^{N} \frac{\alpha_{k,n,t}(2^{r_{k,n,t}} - 1)(\sigma^2 + I_{k,n,t})}{g_{k,n,t}} \right)$$

$$+ \lambda_k \left(Br_k - w\tau \sum_{t=1}^{T} \sum_{n=1}^{N} \alpha_{k,n,t} r_{k,n,t} \right)$$

$$+ \mu_{k,t} \left(P_k^{\max} g_{k,n,t} - \sum_{n=1}^{N} \alpha_{k,n,t}(2^{r_{k,n,t}} - 1)(\sigma^2 + I_{k,n,t}) \right)$$

$$+ \delta_{k,n,t} \alpha_{k,n,t} (\zeta - (2^{r_{k,n,t}} - 1)(\sigma^2 + I_{k,n,t})/I_{k,n,t}); \tag{8.14}$$

here, λ_k, $\mu_{k,t}$, and $\delta_{k,n,t}$ are the Lagrange multipliers being associated with constraints C_1, C_2, and C_3, respectively.

By solving $\nabla \mathcal{L}(r_{k,t}, \lambda_k, \mu_{k,t}, \delta_{k,n,t}) = 0$, and ensuring that all the Karush–Kuhn–Tucker (KKT) conditions [30] are met, we obtain [24]:

$$r_{k,n,t} = \max \left(0, \log_2 \chi + \log_2 \left(\frac{wg_{k,n,t}}{\ln(2)(\sigma^2 + I_{k,n,t})} \right) \right), \tag{8.15}$$

where χ can be represented as

$$\chi = \frac{\lambda_k^{\star}}{(1/T - (\mu_{k,t}^{\star} + \delta_{k,n,t}^{\star} g_{k,n,t}/I_{k,n,t})/\tau)}. \tag{8.16}$$

Equation (8.16) is a form of water-filling solution that may be analyzed using iterative methods, such as Secant and Newton–Raphson [35]. The power allocation is then carried out in two stages. To begin, the transmit power of each user is distributed evenly among its Sk subcarriers by averaging it across the maximum power level of each UT. This is then used to compute the initial interference levels for each group of users that share the same subcarrier. The interference is determined by the SIC process, and the result is determined by the reference threshold ζ (see Equation (8.3)). As it is mentioned in Section 8.2.1, the SIC process at the BS would decode the user with the best channel gain first (when assuming equal

---

Algorithm 8.2: Power optimization for low EMF exposure. Source: Jamshed et al. [24]/with permission of IEEE.

1: **INPUT** $(\alpha, S, K, \zeta, SAR_k, P_k^{\max}, Br_k, \tau, \bar{p}_k(T), \sigma^2, w)$
2: Set $p_{k,n,t} = P_k^{\max}/S_k \; \forall k$;
3: **for** $k = 1: K$ **do**
4:     Calculate initial $I_{k,n,t}$ by using (8.4) $\forall n, t$;
5: **end for**
6: **repeat**
7:     **for** $k = 1: K$ **do**
8:       Calculate $r_{k,n,t}$ by using (8.15) via iterative water-filling;
9:       Calculate $p_{k,n,t}$ by using (8.12);
10:       **if** $p_{k,n,t}g_{k,n,t}/I_{k,n,t} < \zeta$ **then**
11:         Find $r_{k,n,t}$ and $p_{k,n,t}$ that satisfy $p_{k,n,t}g_{k,n,t}/I_{k,n,t} \geq \zeta$;
12:       **end if**
13:       Re-calculate $I_{k,n,t}$ by using (8.4);
14:       Calculate E_k by using (8.6);
15:     **end for**
16: **until convergence**
17: Calculate E by using (8.9);
18: **OUTPUT** E

---

transmit power), i.e. in descending order: $g_{\pi(1),n,t} \geq g_{\pi(2),n,t}, \geq \cdots, \geq g_{\pi(K_{n,t}),n,t}$ such that the user with the worst channel gain will be decoded as free from interference if C_3 in (8.13), which is dependent of ζ, is always met. Second, iterative water-filling is used to optimize the power levels of each k user who is assigned to the same subcarrier n while fulfilling the QoS constraint Br_k and transmit power constraint $P_k^{\max}$. The power level must also satisfy constraint C_3, and if it does not, it is recomputed. The two-stage procedure is then continued until convergence, that is, when the various values of $p_{k,n,t}$ between two iterations of this method stay substantially similar. More information on the power allocation algorithm may be found in Algorithm 8.2.

8.5 Numerical Analysis

This section uses MATLAB simulations to verify the effectiveness of our suggested "ML"-based subcarrier allocation and power optimization for a "PD-NOMA" system. We assumed a single cell of radius R, with the BS in the center (as shown in Figure 8.1) and K users distributed randomly inside it. We studied the propagation effects between any user and the BS using the path-loss model provided in [36]

Table 8.2 Default simulation parameters.

Symbol	Meaning
R	500 m
P_0	−112 dBm
τ	1 ms
P^{ref}	1 W
K	15
W	10 MHz
T	10
N	128
$P_k^{\max}$	0.2 W
σ^2	−174 dBm/Hz
SAR_k	1 W/kg
SAR_k^a	0.658545 W/kg
SAR_k^b	0.561139 W/kg
a	4 bits

Source: Jamshed et al. [24]/with permission of IEEE.

(see page 378) in combination with Rayleigh fading. Table 8.2 lists the exact simulation parameters. We assumed that all users in the cell use comparable devices that adhere to the Federal communications commission (FCC) rules; therefore SAR_k and P^{ref} are constants. We also assumed that all users had the same desired bit count, i.e. $Br_1 = Br_2 = \cdots = Br_K$. Furthermore, we have put $\zeta = 1$ in our simulation. In our simulation scenario, we first utilize Algorithm 8.1 to categorize users and allocate subcarriers before sending the output to Algorithm 8.2, which dynamically adjusts the transmit power of each subcarrier in a low-exposure manner.

8.5.1 Simulation Results

We compared our proposed approach with the offline scheme of [14], the spectrum efficient strategy of [18], and the greedy scheme in Figure 8.3 by adjusting the number of target bits for $T = 10$ time slots, $K = 15$ users, and $N = 128$ subcarriers. It should be emphasized that the criteria for subcarrier allocation in the greedy scheme are based on the user having the best channel characteristics on each subcarrier per time slot. It is obvious that the overall EMF-exposure for uplink communication grows as the goal number of bits increases, because each user requires more power to transmit more information for a fixed T and N value. In comparison with [14] conventional EMF-aware OFDMA approach, our

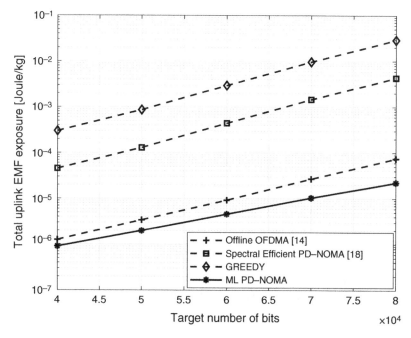

Figure 8.3 The comparison of total EMF uplink exposure vs. the target number of bits for a fixed $K = 15$ users, $T = 10$ time slots, and $N = 128$ subcarriers. Source: Jamshed et al. [24]/with permission of IEEE.

suggested ML PD-NOMA based strategy achieves an exposure reduction of 28% for $Br = 40$ kbit, and even 50% for $Br = 60$ kbit. This reduction in EMF exposure may be explained by the fact that our method, which employs PD-NOMA, can enable a maximum of $K_{n,t}$ users to use the same subcarrier, thereby boosting spectral efficiency when compared with [14] OFDMA equivalent. Furthermore, using SIC as a decoding approach at the receiver end helps to decrease total uplink EMF exposure by enabling the user with the worst channel gain to broadcast data free of interference, hence decreasing overall transmit power. When compared with the spectrum efficient PD-NOMA of [18] and the greedy scheme, our suggested ML PD-NOMA-based technique lowers EMF exposure by at least 2 and 3 orders of magnitude for $Br = 40$ kbit. When comparing [18] to our suggested ML PD-NOMA based technique and [14], an increase in EMF exposure is noticed as the system strives to optimize the weighted sum-rate of K users over each t time slot. Furthermore, the greedy strategy has the highest overall EMF exposure since it uses all of the available transmit power on each subcarrier in an attempt to increase the number of bits transferred.

To confirm the effectiveness of our proposed scheme in comparison with the state-of-the-art techniques in [14, 18], and the greedy scheme, we compare these four schemes in Figure 8.4 for a varying value of K when the target number of bits is fixed to 60 or 80 kbit. When the number of subcarriers and the transmission window are unchanged, the total EMF exposure for uplink communication rises with the number of users because more transmission power is required for each user to send the same amount of information. As a result, as the value of S_k depends on the number of users, the number of subcarriers assigned to each user reduces as the number of users rises, and therefore the total EMF exposure increases. In comparison with [14], our scheme exhibits a consistent trend, as PD-NOMA permits each subcarrier to be shared by a maximum of $K_{n,t}$, resulting in a decrease of at least 25% in total EMF exposure for a fixed $Br = 60$ kbit. When $Br = 80$ kbit, our suggested strategy minimizes EMF exposure by at least 38% when compared with [14] offline scheme. Furthermore, when comparing the suggested ML PD-NOMA-based approach with the spectrum efficient PD-NOMA of [18] and the greedy scheme, our method decreases EMF-exposure by at least two and three orders of magnitude for $Br = 60$ kbit. It should also be noted that the performance disparity between the schemes widens as the number of users grows. With example, our design can handle four more people (22 instead of 18) as [14] scheme for the same total exposure of 3×10^{-5} J/kg.

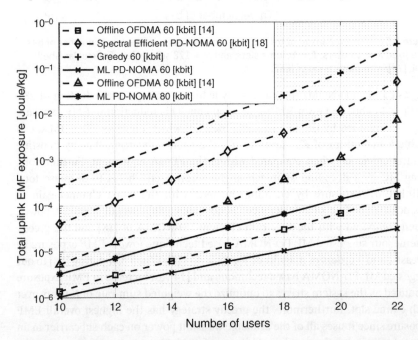

Figure 8.4 The comparison of total EMF uplink exposure vs. varying the number of user for different target number of bits and for a fixed $T = 10$ time slots. Source: Jamshed et al. [24]/with permission of IEEE.

In Figure 8.5, we compare the performance of our scheme to that of [14, 18], and the greedy scheme, for varied numbers of time slots and by setting the values of K and N, where the goal number of bits is either 60 or 80 kbit. According to Figure 8.5, increasing the number of time slots reduces EMF exposure, which might be attributed to the fact that each user can lower transmission power by dispersing the desired number of bits throughout the growing value of T. Furthermore, as compared with [14] offline system, our scheme minimizes overall EMF exposure. In comparison with the offline method of [14], our suggested scheme decreases exposure by at least 21% and 34% for fixed values of $Br = 60$ kbit and $Br = 80$ kbit, respectively. This performance boost can be attributed to the fact that resources are more readily available in PD-NOMA than in OFDMA. Although sharing a subcarrier by $K_{n,t}$ users introduces extra interference in addition to noise, the adoption of SIC as a decoding approach at the receiver end allows it to be mitigated. In comparison with the [18] and greedy schemes, our strategy decreases EMF-exposure by at least one and two orders of magnitude for $Br = 60$ kbit, respectively. To confirm the change of the total uplink EMF exposure on increasing the number of time slots over variable target number of bits, we

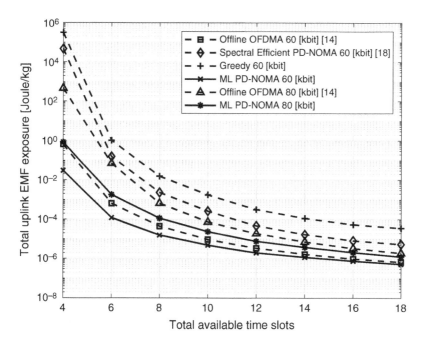

Figure 8.5 The comparison of total EMF uplink exposure vs. varying the time slots for different target number of bits and for a fixed $K = 15$ users. Source: Jamshed et al. [24]/with permission of IEEE.

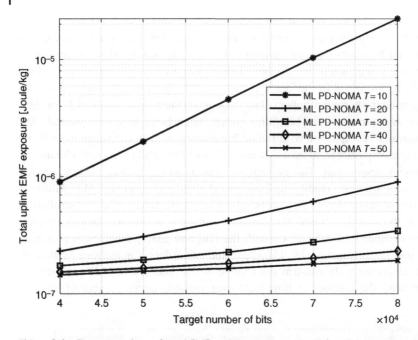

Figure 8.6 The comparison of total EMF uplink exposure vs. varying the target number of bits for different values of T and for a fixed $K = 15$ users. Source: Jamshed et al. [24]/with permission of IEEE.

evaluated the impact of increasing the value of T on the total EMF exposure for a fixed value of K and N in Figure 8.6. It is clear and confirms our findings in Figure 8.5 that as the number of time slots increases, the EMF exposure reduces, which may be attributed to the fact that each user can lower transmission power by dispersing the intended number of bits over an increasing value of T.

8.5.2 Scheme Validity for Real Applications

In this part, we used simulation findings to compute the value of the SAR for a PIFA with one transmit antenna accessible for uplink transmission and radiating at a resonance frequency of 2.1 GHz. We used the Voxel models (heterogeneous human biological replication) available in computer simulation technology (CST) 2018 to determine the maximum SAR value for 1g of body mass, according per IEEE guidelines [37]. Figure 8.7 depicts these two scenarios: (a) the cheek position and (b) the tilt position. We measured the highest SAR value using the IEEE/IEC 62704-1 averaging technique cite8088404 and the maximum transmit power $P_k^{\max} = 0.2$ W. The SAR values produced from these simulations are provided in Table 8.2 as SAR_k^a and SAR_k^b. We exhibit the change of total uplink EMF exposure as a function of bit number in Figure 8.8 by utilizing SAR_k^a (for cheek position),

Figure 8.7 Illustration of SAR testing positions (a) Cheek position, (b) Tilt position [37]. Source: Jamshed et al. [24]/with permission of IEEE.

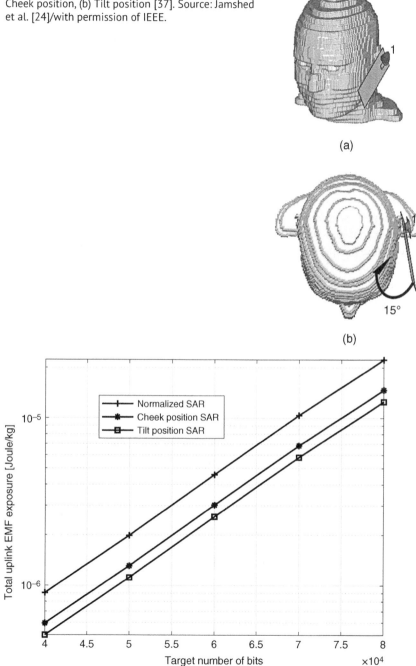

(a)

(b)

Figure 8.8 Comparison of total uplink exposure with target number of bits for different SAR values. Source: Jamshed et al. [24]/with permission of IEEE.

SAR_k^b (for tilt position), and normalized SAR_k values. The results in Figure 8.8 demonstrate a similar pattern for three situations, however the values of SAR obtained via experimental analysis result in a reduction in total uplink EMF exposure, which may be supported by the lower predicted maximum SAR values for the two scenarios. Furthermore, the low SAR values strengthen the case for utilizing an acPIFA in mobile devices.

8.6 Conclusion

In this chapter, we presented an ML-based user grouping and resource allocation technique to reduce EMF exposure for PD-NOMA systems' uplink. To begin, we intelligently aggregate users into different clusters and execute subcarrier allocation using the ML-based user grouping and resource allocation method. Second, we execute power assignment utilizing the assigned subcarriers by exploiting a change in variables to build a convex optimization problem. In comparison with state-of-the-art approaches, our strategy significantly minimizes EMF exposure. Furthermore, we have provided the SAR values, which were determined using simulation scenarios based on IEEE standards. We demonstrated that our approach can still minimize overall EMF exposure by employing an PIFA with a single antenna, and we further validated the feasibility of using an PIFA in mobile devices. In the future, we plan to expand our research to include multi-antenna systems.

References

1 Boccardi, F., Heath, R.W., Lozano, A. et al. (2014). Five disruptive technology directions for 5G. *IEEE Communications Magazine* 52 (2): 74–80.

2 International Agency for Research on Cancer and others (2011). IARC classifies radiofrequency electromagnetic fields as possibly carcinogenic to humans. *Press release* no. 208.

3 Hardell, L.O., Carlberg, M., Söderqvist, F. et al. (2007). Long-term use of cellular phones and brain tumours-increased risk associated with use for >10 years. *Occupational and Environmental Medicine* 64 (9): 626–632.

4 Radiofrequency Electromagnetic Fields (1997). Evaluating compliance with FCC guidelines for human exposure to radio frequency electromagnetic fields. *OET Bull* 65: 1–53.

5 ICNIRP. Guidelines for Limiting Exposure to Time-Varying Electric, Magnetic and Electromagnetic Fields (100 kHz to 300 GHz). https://www.icnirp.org/cms/upload/consultation-upload/ICNIRP-RF-Guidelines-PCD-2018-07-11.pdf (accessed 26 February 2019).

6 Federal Communications Commission (2014). Specific Absorption Rate (SAR) for Cell Phones: What it Means for You.

7 Dai, L., Wang, B., Ding, Z. et al. (2018). A survey of non-orthogonal multiple access for 5G. *IEEE Communication Surveys and Tutorials* 20 (3): 2294–2323.

8 Mounchili, S. and Hamouda, S. (2020). Pairing distance resolution and power control for massive connectivity improvement in NOMA systems. *IEEE Transactions on Vehicular Technology* 69 (4): 4093–4103.

9 Cui, J., Ding, Z., Fan, P., and Al-Dhahir, N. (2018). Unsupervised machine learning-based user clustering in millimeter-wave-NOMA systems. *IEEE Transactions on Wireless Communications* 17 (11): 7425–7440.

10 Nikopour, H., Yi, E., Bayesteh, A. et al. (2014). SCMA for downlink multiple access of 5G wireless networks. *2014 IEEE Global Communications Conference*, pp. 3940–3945, IEEE.

11 Cai, D., Fan, P., and Mathiopoulos, P.T. (2017). A tight lower bound for the symbol error performance of the uplink sparse code multiple access. *IEEE Wireless Communications Letters* 6 (2): 190–193.

12 Jamshed, M.A., Heliot, F., and Brown, T. (2019). A survey on electromagnetic risk assessment and evaluation mechanism for future wireless communication systems. *IEEE Journal of Electromagnetics, RF and Microwaves in Medicine and Biology* 4 (1): 24–36.

13 Islam, M.T., Faruque, M.R.I., and Misran, N. (2009). Reduction of specific absorption rate (SAR) in the human head with ferrite material and metamaterial. *Progress in Electromagnetics Research* 9: 47–58.

14 Sambo, Y.A., Al-Imari, M., Héliot, F., and Imran, M.A. (2016). Electromagnetic emission-aware schedulers for the uplink of OFDM wireless communication systems. *IEEE Transactions on Vehicular Technology* 66 (2): 1313–1323.

15 Sambo, Y.A., Heliot, F., and Imran, M.A. (2017). Electromagnetic emission-aware scheduling for the uplink of multicell OFDM wireless systems. *IEEE Transactions on Vehicular Technology* 66 (9): 8212–8222.

16 De Domenico, A., Díez, L.F., Agüero, R. et al. (2015). EMF-aware cell selection in heterogeneous cellular networks. *IEEE Communications Letters* 19 (2): 271–274.

17 Zuo, H. and Tao, X. (2017). Power allocation optimization for uplink non-orthogonal multiple access systems. *2017 9th International Conference on Wireless Communications and Signal Processing (WCSP)*, pp. 1–5, IEEE.

18 Al-Imari, M., Xiao, P., and Imran, M.A. (2015). Receiver and resource allocation optimization for uplink NOMA in 5G wireless networks. *2015 International Symposium on Wireless Communication Systems (ISWCS)*, pp. 151–155, IEEE.

19 Yin, Y., Peng, Y., Liu, M. et al. (2019). Dynamic user grouping-based NOMA over Rayleigh fading channels. *IEEE Access* 7: 110964–110971.

20 Jain, A.K. (2010). Data clustering: 50 years beyond K-means. *Pattern Recognition Letters* 31 (8): 651–666.

21 Bindra, K., Mishra, A., and Suryakant (2019). Effective data clustering algorithms. In: *Soft Computing: Theories and Applications*, 419–432. Springer.

22 Di, B., Song, L., and Li, Y. (2016). Sub-channel assignment, power allocation, and user scheduling for non-orthogonal multiple access networks. *IEEE Transactions on Wireless Communications* 15 (11): 7686–7698.

23 Jamshed, M.A., Amjad, O., Heliot, F., and Brown, T. (2019). EMF-reduction uplink resource allocation scheme for non-orthogonal multiple access systems. *2019 IEEE Wireless Communications and Networking Conference Workshop (WCNCW)*, pp. 1–5, IEEE.

24 Jamshed, M.A., Heliot, F., and Brown, T.W. (2021). Unsupervised learning based emission-aware uplink resource allocation scheme for non-orthogonal multiple access systems. *IEEE Transactions on Vehicular Technology* 70 (8): 7681–7691.

25 Golden, G.D., Foschini, C., Valenzuela, R.A., and Wolniansky, P.W. (1999). Detection algorithm and initial laboratory results using V-BLAST space-time communication architecture. *Electronics Letters* 35 (1): 14–16.

26 Ali, Z., Sidhu, G.A.S., Waqas, M., and Gao, F. (2018). On fair power optimization in nonorthogonal multiple access multiuser networks. *Transactions on Emerging Telecommunications Technologies* 29 (12): e3540.

27 Conil, E. (2013). D2. 4 Global Wireless Exposure Metric Definition V1. *LexNet Project*.

28 Federal Communications Commission (2001). Evaluating compliance with FCC guidelines for human exposure to radiofrequency electromagnetic fields, supplement C *OET Bulletin* 65: 1–57.

29 3GPP (2015). 3GPP TS 36.213, 'Evolved Universal Terrestrial Radio Access (E-UTRA); Physical layer procedures (Release 12).

30 Luo, Z.-Q. and Yu, W. (2006). An introduction to convex optimization for communications and signal processing. *IEEE Journal on Selected Areas in Communications* 24 (8): 1426–1438.

31 Al-Imari, M., Xiao, P., Imran, M.A., and Tafazolli, R. (2013). Low complexity subcarrier and power allocation algorithm for uplink OFDMA systems. *EURASIP Journal on Wireless Communications and Networking* 2013 (1): 1–6.

32 Ye, H., Li, G.Y., and Juang, B.-H.F. (2019). Deep reinforcement learning based resource allocation for V2V communications. *IEEE Transactions on Vehicular Technology* 68 (4): 3163–3173.

33 Tibshirani, R., Walther, G., and Hastie, T. (2001). Estimating the number of clusters in a data set via the gap statistic. *Journal of the Royal Statistical Society: Series B (Statistical Methodology)* 63 (2): 411–423.

34 Goutte, C., Toft, P., Rostrup, E. et al. (1999). On clustering FMRI time series. *NeuroImage* 9 (3): 298–310.

35 Abramowitz, M. and Stegun, I. (1964). *Handbook of Mathematical Functions: With Formulas, Graphs, and Mathematical Tables.* Applied mathematics series. National Bureau of Standards, Washington, DC.

36 Wu, J., Rangan, S., and Zhang, H. (2016). *Green Communications: Theoretical Fundamentals, Algorithms, and Applications.* CRC Press.

37 IEEE Std 1528-2013 (2013). *IEEE Recommended Practice for Determining the Peak Spatial-Average Specific Absorption Rate (SAR) in the Human Head from Wireless Communications Devices: Measurement Techniques (Revision of IEEE Std 1528-2003)*, pp. 1–246.

9

Emission-Aware Resource Optimization for Backscatter-Enabled NOMA Networks

Muhammad Ali Jamshed[1], Wali Ullah Khan[2], Haris Pervaiz[3],
Muhammad Ali Imran[1], and Masood Ur Rehman[1]

[1]*James Watt School of Engineering, University of Glasgow, Glasgow, UK*
[2]*Interdisciplinary Center for Security, Reliability and Trust (SnT), University of Luxembourg, Luxembourg City, Luxembourg*
[3]*School of Computing and Communications, Lancaster University, Lancaster, UK*

9.1   Introduction

The rapid growth of wireless communication has increased the number of user proximity wireless devices (UPWDs). To accommodate this increase in UPWDs, capacity has to grow by 1,000 folds. This capacity expansion will partially be met by densifying the wireless infrastructure [1]. This densification will increase the levels of EM field (EMF) exposure significantly. Although, the literature does not provide strong evidence of severe short term impacts of EMF on human health, the international agency for research on cancer (IARC) and world health organization (WHO) have classified the EMF waves from UPWDs as carcinogenic to humans [2]. Moreover, the upsurge in UPWDs could possibly increase the long term health hazards as well [3]. To minimize the health effects of EMF, the UPWDs must comply with by International Commission on Non-Ionizing Radiation Protection (ICNIRP) (at International level) and federal communications commission (FCC) (in the USA) exposure limits and regulations [4]. However, it is interesting to note that FCC consumer guidelines state; "FCC approval does not indicate the amount of EMF exposure consumers experience during normal use of the device" [5].

Use of non-orthogonal multiple access (NOMA) in modern wireless communication networks has proven to be an effective multiplexing strategy for augmented spectral efficiency and massive connectivity. There are two variants of NOMA, i.e. code-domain NOMA (CD-NOMA) and power domain-NOMA (PD-NOMA) and for the scope of this chapter, we have focused on the latter [6]. In PD-NOMA,

Low Electromagnetic Field Exposure Wireless Devices: Fundamentals and Recent Advances, First Edition.
Edited by Masood Ur Rehman and Muhammad Ali Jamshed.

the UPWDs having a difference in channel properties are allocated to the same sub-carrier. The UPWDs sharing a similar resource block are multiplexed based on different power levels by employing superposition coding technique at the transmitter side and are decoded using successive interference cancellation (SIC) at the receiver end [7].

The upsurge in data requirements has significantly increased the energy needs of UPWDs that directly impact the amount of EMF absorbed by a UPWDs user. The use of ambient backscatter communications (ABC) has emerged as a promising solution to overcome the issues related to energy consumption [8]. The ABC uses existing radio frequency (RF) signals and can provide a path for data transmission between the source and the destination. The ABC can reflect the RF signals towards the intended UPWDs without altering any oscillatory circuity, hence, maximizing the energy efficiency of UPWDs [9]. These classical properties of PD-NOMA and ABC provide an effective road map to reduce the amount of EMF absorbed by UPWDs users, while maintaining quality of service (QoS).

9.1.1 Motivation and Contributions

A limited amount of work is available in the literature on reducing the amount of EMF in the uplink of a wireless system. In [10], a heuristic approach is adopted to reduce the EMF in the uplink of orthogonal frequency division multiple access (OFDMA) based cellular system, while incorporating the classical definition of exposure dose metric (EMF over time). In [11] the similar dose metric is utilized, to propose a machine learning (ML)-based uplink resource allocation scheme to reduce the EMF in the uplink of PD-NOMA system. In this work, a new scheduling framework that relies on ABC, PD-NOMA, and ML technologies to reduce the EMF in the uplink of wireless systems is proposed. In comparison with other well-known EMF reduction strategies, the proposed framework reduces the EMF exposure by an equitable percentage. The ABC further improves the channel gain between UPWDs and the base station (BS), hence helping to further reduce the EMF. The k-medoids is used to group UPWDs into different clusters, where the number of clusters is found using Silhouette analysis. To the best of authors' knowledge, this is the first work that have relied on ABC, PD-NOMA, and ML to minimize the uplink EMF. The contributions of this study are summarized as follows:

1. We have formulated a PD-NOMA based multi-user EMF scheduling scheme. In comparison with [11], we have incorporated ABC, which makes the design more challenging but provides an opportunity to achieve higher reductions in the EMF.
2. In comparison with [11], we have used k-medoids instead of k-means for the user grouping. In comparison with k-means, k-medoids is more robust, less

complex, and takes less time to converge [12]. Moreover, in comparison with [11], and instead of relying on multiple methods, i.e. F-test and elbow method, we have used Silhouette analysis to find the number of users per sub-carrier. Lastly, the power allocation is carried out in a EMF-aware manner.

3. The Monte Carlo types of simulations are used to validate the performance of proposed optimization framework. In comparison with [10] and [11], the proposed ABC, PD-NOMA, and ML based optimization framework reduces the EMF by at least 82 and 75%, respectively.

The remainder of the chapter is organized as follows: the system model and the problem formulation is provided in Section 9.2, which shows mathematical framework and the scenario considered to design the ABC and PD-NOMA based EMF-aware technique. Section 9.3 explains the proposed EMF-aware resource and power allocation strategies. The simulation carried out to study the superiority of our proposed technique over the similar strategies is provided in Section 4.2. Finally, the conclusion are drawn in Section 9.5.

9.2 System Model

A single cell scenario of PD-NOMA and ABC based communication system is shown in Figure 9.1. The single cell of radius R_s is equipped with a single BS and U UPWDs incorporated with a single antenna and are able to communicate with the BS. To facilitate the communication between UPWDs and the BS, b, backscatter tags are available within the coverage area of BS. The system bandwidth W is equally divide into O sub-carriers such that, $w = W/|O|$. Based on NOMA property, $U_{o,\hat{t}}$ users can be allocated to a o sub-carrier at a given $\hat{t}$ time slot, where $U_{o,\hat{t}} > 1$. Since, we have considered an uplink scenario, a backscatter tag b also receives RF signal over o sub-carrier, harvest its own energy, modulate its own information, and reflect it towards BS. Thus, the received signal at the BS can be expressed as

$$
y_{o,\hat{t}} = \sum_{u=1}^{U_{o,\hat{t}}} \left(\sqrt{p_{u,o,\hat{t}}\, g_{u,o,\hat{t}}}\, x_{u,o,\hat{t}} \right.
$$
$$
\left. + \sqrt{p_{u,o,\hat{t}}\, \xi_b g_{b,o,\hat{t}}^u g_{u,o,\hat{t}}^b}\, z_{u,o,\hat{t}} x_{u,o,\hat{t}} \right) + \omega_{o,\hat{t}}, \tag{9.1}
$$

here $p_{u,\hat{t}}$ represents the transmit power of the uth UPWD at $\hat{t}$ time slot and o sub-carrier, $g_{k,o,\hat{t}}$ represents the channel gain between the BS and the uth UPWD at $\hat{t}$ time slot and o sub-carrier, $x_{u,o,\hat{t}}$ is the information signal of uth UPWD at $\hat{t}$ time slot and o sub-carrier. Accordingly, ξ_b is the reflection coefficient of backscatter b, $g_{b,o,\hat{t}}^u$ represents the channel gain between u^{th} UPWD and backscatter b at $\hat{t}$ time slot and o sub-carrier, $g_{u,o,\hat{t}}^b$ denotes the channel gain between backscatter tag b and BS at $\hat{t}$

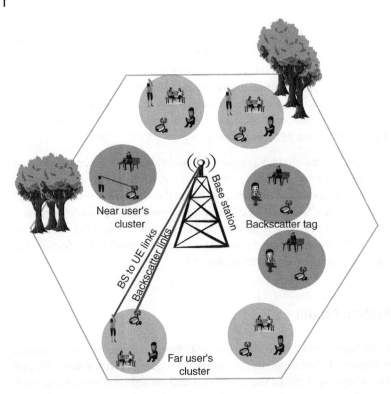

Figure 9.1 Illustration of the system model.

time slot and o sub-carrier, and $\omega_{o,\hat{t}}$ is the additive white Gaussian noise (AWGN) with a zero mean and variance of σ^2, at $\hat{t}$ time slot and o sub-carrier. In NOMA, the interference occurs due to the multiplexing of different users data on a single sub-carrier. The uth user multiplexed on sub-carrier o experiences the following total interference:

$$\bar{I}_{u,o,\hat{t}} = \sum_{l=1,l\neq u}^{U_{o,\hat{t}}} p_{l,o,\hat{t}}(g_{l,o,\hat{t}} + \xi_b g_{b,o,\hat{t}}^l g_{l,o,\hat{t}}^b). \tag{9.2}$$

In order to perform decoding of the users multiplexed together, the SIC is executed at the signal receiving side. In the uplink scenario of PD-NOMA, firstly, the users with best channel gain are demultiplexed, which is then followed by decoding of the worst channel gain users [13]. In PD-NOMA, the multiplexing strategy highly depends on the ability of SIC to successfully decode the multiplexed signal [14]. This can be achieved if the ratio of received signal to residual interference of user u on sub-carrier o of $\hat{t}$ time slot meets the following condition:

$$p_{u,o,\hat{t}}(g_{u,o,\hat{t}} + \xi_b g_{b,o,\hat{t}}^u g_{u,o,\hat{t}}^b)/I_{u,o,\hat{t}} \geq \zeta, \tag{9.3}$$

the reference threshold $\zeta \geq 1$, where

$$I_{u,o,\hat{t}} = \sum_{l=\pi^{-1}(u)+1}^{U_{o,\hat{t}}} p_{\pi(l),o,\hat{t}} \left(g_{\pi(l),o,\hat{t}} + \xi_b g_{b,o,\hat{t}}^{\pi(l)} g_{\pi(l),o,\hat{t}}^b \right). \tag{9.4}$$

Using the Shannon formula, the transmitted number of bits over a τ duration of time slot $\hat{t}$ and sub-carrier o by a user u is expressed as

$$bt_{u,o,\hat{t}(\alpha,p)} = w\tau \, \alpha_{u,o,\hat{t}} \log_2 \left(1 + \frac{p_{u,o,\hat{t}}(g_{u,o,\hat{t}} + \xi_b g_{b,o,\hat{t}}^u g_{u,o,\hat{t}}^b)}{\sigma^2 + I_{u,o,\hat{t}}} \right), \tag{9.5}$$

where, $\alpha_{u,o,\hat{t}}$ is sub-carrier allocation index and w is the bandwidth of each sub-carrier. The level of EMF in the uplink of a single user u is defined as [15]

$$E_u(\alpha,p) = \frac{SAR_u}{\hat{T} P^{\text{ref}}} \tau \left(\hat{p}_u(\hat{T}) + \sum_{\hat{t}=1}^{\hat{T}} \sum_{o=1}^{O} \alpha_{u,o,\hat{t}} p_{u,o,\hat{t}} \right), \tag{9.6}$$

here, P^{ref} is reference incident power used to estimate specific absorption rate SAR, $\hat{T}$ is total number of time slots, and $\hat{p}_u(\hat{T})$ is the signaling power. The SAR and $\hat{p}_u(\hat{T})$ can be computed as discussed in [3].

9.2.1 Problem Formulation

In this work, the objective is to reduce the total uplink EMF while considering a PD-NOMA and ABC enabled cellular framework. The optimization problem is formulated in the form of mathematical expressions and is defined as follows:

$$(OP) \min_{p,\alpha} \quad E(\alpha,p) = \sum_{u=1}^{U} E_u(\alpha,p),$$

s.t:

$$(C1): \quad \sum_{\hat{t}=1}^{\hat{T}} \sum_{o=1}^{O} bt_{u,o,\hat{t}}(\alpha,p) = Bt_u, \forall u$$

$$(C2): \quad \sum_{o=1}^{O} \alpha_{u,o,\hat{t}} p_{u,o,\hat{t}} \leq P_u^{\max} \quad \forall u, \forall \hat{t},$$

$$(C3): \quad \alpha_{u,o,\hat{t}}(p_{u,o,\hat{t}}(g_{u,o,\hat{t}} + \xi_b g_{b,o,\hat{t}}^u g_{u,o,\hat{t}}^b))/I_{u,o,\hat{t}} \geq \zeta \quad \forall u, \forall o, \forall \hat{t},$$

$$(C4): \quad \sum_{u=1}^{U} \alpha_{u,o,\hat{t}} \leq U_{o,\hat{t}} \quad \forall o, \forall \hat{t},$$

$$(C5): 0 \leq \xi_b \leq 1 \quad \forall b, \tag{9.7}$$

where, the total EMF absorbed by U users over $\hat{T}$ time slots is expressed by objective function $E(\alpha,p)$ in (OP). The constraint (C1) ensures the QoS of each user.

The (C2) constraint is related to the maximum power limit $P_u^{\max}$ of each user. The (C3) constraint is related to the successful implementation of SIC. The constraint (C4) reflects the maximum allowable users on a single sub-carrier. Finally, the constraint (C5) controls the reflection of backscatter tags. The binary nature of $\alpha_{u,o,\hat{i}}$ and non-affine nature the (C1) constraint makes the problem non-convex [16]. A standard relaxation procedure needs to be applied to overcome the binary nature of $\alpha_{u,o,\hat{i}}$. This is catered by adopting a sequential sub-carrier and power allocation strategy, i.e. performing sub-carrier allocation for a fixed power and vice versa [17]. Moreover, it is assumed that the channel state information (CSI) is available by exploiting uplink pilot signals.

9.3 Proposed Solution

9.3.1 Sub-carrier Allocation

In PD-NOMA the $U_{o,\hat{i}}$ users can be assigned to a single sub-carrier by exploiting the variations in their channel properties and are demultiplexed by using the levels of received power. The sub-carrier assignment relies on receiver capability to differentiate between power levels, which itself relies on grouping of users on same sub-carrier. In comparison with heuristic techniques, the use of clustering based ML algorithms have proven to be much effective, less complex and have a higher probability of convergence [12]. Here, similar to [11], we have used ML techniques to perform the sub-carrier allocation. However, instead of using k-means, the k-medoids based clustering is employed due to its low complexity and robustness. Moreover, instead of relying on traditional F-test and elbow methods for predicting the best number of clusters, a Silhouette analysis has been carried out. The elbow method, provides a range based on elbow criteria, hence creating ambiguities in selecting the best value of C. Whereas, the Silhouette analysis is more robust and removes such ambiguities.

Figure 9.2 shows the trend of Silhouette values by varying C, for $U = 100, O = 1$, and $\hat{T} = 1$ over a Rayleigh fading channel and path loss model defined in [18]. The clusters are formed using k-medoids. It can be observed that for $C = 2$, a Silhouette value of 1 is achieved, which corresponds to number of users that can be multiplexed on a single sub-carrier. Once a suitable value of C is selected, the sub-carrier allocation is performed by relying on normalized channel gain, i.e. $G_{u,o,\hat{i}}$, to maintain fairness among all users.

9.3.2 Power Allocation

After the learning-based sub-carrier allocation, the power for each user needs to be allocated optimally to minimize the uplink EMF. The constraint (C1) in (OP)

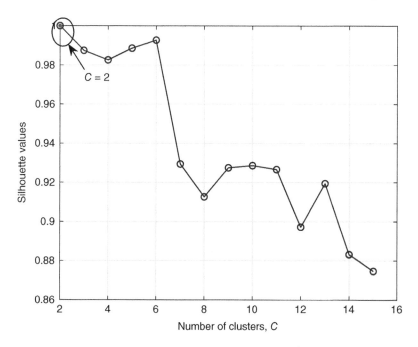

Figure 9.2 Silhouette values vs. C for $O = 1$, $U = 100$, and $\hat{T} = 1$.

is still non-convex and to overcome its non-linear nature, the (9.5) is transformed as follows:

$$p_{u,o,\hat{i}} = \frac{(2^{r_{u,o,\hat{i}}} - 1)(\sigma^2 + I_{u,o,\hat{i}})}{g_{u,o,\hat{i}} + \xi_b g^u_{b,o,\hat{i}} g^b_{u,o,\hat{i}}}, \tag{9.8}$$

and by using $p_{u,o,\hat{i}}$, the (OP) can be re-defined as follows:

$$\min_r \frac{\tau}{\hat{T}} \sum_{u=1}^{U} \left(\hat{p}_u(\hat{T}) + \sum_{\hat{i}=1}^{\hat{T}} \sum_{o=1}^{O} \frac{\alpha_{u,o,\hat{i}}(2^{r_{u,o,\hat{i}}} - 1)(\sigma^2 + I_{u,o,\hat{i}})}{(g_{u,o,\hat{i}} + \xi_b g^u_{b,o,\hat{i}} g^b_{u,o,\hat{i}})} \right),$$

s.t:

$$(C6): \quad w\tau \sum_{\hat{i}=1}^{\hat{T}} \sum_{o=1}^{O} \alpha_{u,o,\hat{i}} r_{u,o,\hat{i}} = Bt_u,$$

$$(C7): \quad \sum_{o=1}^{O} \frac{\alpha_{u,o,\hat{i}}(2^{r_{u,o,\hat{i}}} - 1)(\sigma^2 + I_{u,o,\hat{i}})}{(g_{u,o,\hat{i}} + \xi_b g^u_{b,o,\hat{i}} g^b_{u,o,\hat{i}})} \leq P_u^{\max},$$

$$(C8): \quad \alpha_{u,o,\hat{i}}((2^{r_{u,o,\hat{i}}} - 1)(\sigma^2 + I_{u,o,\hat{i}})/I_{u,o,\hat{i}}) \geq \zeta, \tag{9.9}$$

here, the spectral efficiency of u user is defined as $r_{u,o,\hat{\imath}} = bt_{u,o,\hat{\imath}}/(w\tau)$. In comparison with (OP), (9.9), as a function of $\mathbf{r}$ is affine. Since, the problem defined in (9.9) is convex, its Lagrangian can be found easily, defined as follows:

$$r_{u,o,\hat{\imath}} = \max\left(0, \log_2 \chi + \log_2\left(\frac{w(g_{u,o,\hat{\imath}} + \xi_b g^u_{b,o,\hat{\imath}} g^b_{u,o,\hat{\imath}})}{\ln(2)(\sigma^2 + I_{u,o,\hat{\imath}})}\right)\right), \tag{9.10}$$

where, λ_u, $\mu_{u,\hat{\imath}}$ and $\delta_{u,o,\hat{\imath}}$ are associated Lagrange multipliers of (C6), (C7) and (C8). The χ is expressed as

$$\chi = \frac{\lambda_u^\star}{(1/\hat{T} - (\mu_{u,\hat{\imath}}^\star + \delta_{u,o,\hat{\imath}}^\star (g_{u,o,\hat{\imath}} + \xi_b g^u_{b,o,\hat{\imath}} g^b_{u,o,\hat{\imath}})/I_{u,o,\hat{\imath}})/\tau)}. \tag{9.11}$$

Equation (9.10) is a water-filling based solution and can be solved by employing different iterative solutions [10]. The details about sub-carrier allocation and power assignment schemes are provided in Algorithm 9.1.

Algorithm 9.1: Low EMF learning based (ABC) and (PD-NOMA)-enabled optimization framework.

1: **INPUT** $(U, \hat{T}, O, g_{u,o,\hat{\imath}}, g^u_{b,o,\hat{\imath}}, g^b_{u,o,\hat{\imath}}, \alpha, \zeta, SAR_u, P^{\max}_u, Bt_u, \tau, \hat{p}_u(\hat{T}), \sigma^2, w,$
$\quad \xi_b, b)$
2: **Step 1:** User grouping
3: **for** $C = 2: U - 1$ **do**
4: $\quad$ Use K-mediods to cluster $\mathbf{g}_{o,\hat{\imath}} = [g_{1,o,\hat{\imath}}, ..., g_{U,o,\hat{\imath}}]$;
5: $\quad$ Set $U_{o,\hat{\imath}} = C$ based on Silhouette analysis;
6: **end for**
7: **Step 2:** Sub-carrier allocation
8: Use $G_{u,o,\hat{\imath}} = g_{u,o,\hat{\imath}}/\hat{g}_u$ $\forall u, o, \hat{\imath}$, where $\hat{g}_u$ is the mean channel gain.
9: Use max. of $G_{,o,\hat{\imath}}$ within each cluster for each user.
10: Allocate $S = \frac{O \times \hat{T}}{U}$ sub-carrier to each user.
11: **Step 3:** Power assignment
12: Set $p_{u,o,\hat{\imath}} = P^{\max}_u/S$ to estimate initial interference;
13: **repeat**
14: $\quad$ **for** $u = 1: U$ **do**
15: $\quad\quad$ Estimate $r_{u,o,\hat{\imath}}$ using iterative water-filling and satisfying SIC constraint;
16: $\quad\quad$ Estimate $p_{u,o,\hat{\imath}}$ utilizing (8);
17: $\quad\quad$ Re-evaluate $I_{u,o,\hat{\imath}}$ utilizing (4);
18: $\quad\quad$ Estimate E_u utilizing (6);
19: $\quad$ **end for**
20: **until convergence**
21: Evaluate E utilizing (OP);
22: **OUTPUT** E;

9.4 Performance Evaluation

The performance of the proposed ABC and PD-NOMA enabled scheme is validated using MATLAB simulations. The coverage radius of BS is set as $R_s = 500$ meters, where U users are randomly placed uniformly. The Rayleigh fading and the path loss model defined in [18] is used to model propagation effects. A single backscatter tag b is considered in each cluster to facilitate the communication between BS and users. We assume that the values of SAR_u and P^{ref} are constants (pertaining to FCC limits), and the number of bits is the same for all users. The $\sigma^2 = -174$ dBm/Hz, $W = 10$ MHz, $\zeta = 1$, $\xi = 1$, $P_{max} = 0.2$ W, $\tau = 1$ ms, $SAR_u = 1$ W/kg, and $P^{ref} = 1$ W are assumed. The sub-carrier and power allocation is performed using the steps defined in Algorithm 9.1.

In Figure 9.3, a comparison of the proposed scheme with [10] and [11], is carried out by increasing the number of bits for $O = 128$ sub-carriers, $U = 15$ users, and $\hat{T} = 10$ time slots. A rising tendency in the EMF can be seen for increasing number of bits as each user requires more transmit power to achieve the required number of bits. In comparison with OFDMA method [10], the proposed ABC and PD-NOMA based scheme achieves 82% reduction in EMF for $Bt = 40$ kbit,

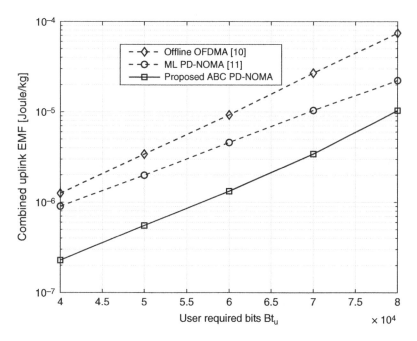

Figure 9.3 Comparing combined uplink EMF against a variation in the bits for $U = 15$, $\hat{T} = 10$, and $O = 128$.

as PD-NOMA possess an increase in spectral efficiency by allowing $U_{o,\hat{i}}$ users to share the same sub-carrier and the use of SIC helps to reduce the uplink power. In comparison with a similar ML-based PD-NOMA strategy [11], our scheme reduces the EMF by 75% for $Bt = 40$ kbit, as the addition of backscatter tags helps to further improve the channel quality between users and the BS.

In Figure 9.4, a comparison of the proposed scheme with the techniques discussed in [10] and [11] is carried out by increasing the value of U for a fixed value of $O = 128$, $\hat{T} = 10$, and $Bt_u = 60$ kbits. An increasing trend in EMF can be seen for larger number of users as each user requires more transmit power to achieve same number of bits, while the number of allocated sub-carrier and the size of transmission window are fixed. In comparison with [10], the proposed ABC and PD-NOMA based scheme allows multiple users to share the same sub-carrier, hence providing at least 87% reduction in EMF. While comparing with the similar ML PD-NOMA scheme of [11], at least 33% reduction in the EMF is achieved, which is largely due to the addition of ABC.

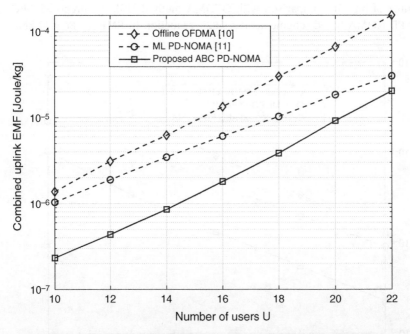

Figure 9.4 Comparing combined uplink EMF against a variation in the number of users for fixed bits, time slots, and sub-carriers.

9.5 Conclusion

A new EMF-aware resource allocation and user grouping scheme is proposed by using ML technologies for ABC and PD-NOMA based wireless systems. The k-medoids and Silhouette analysis has been used to perform user grouping and sub-carrier allocation, which is then followed by power assignment. The power assignment is performed through solving the (OP). In comparison with similar techniques, our proposed framework reduces the EMF by at least 75%. These results indicate the vast potential of this novel method and indicate its suitability to be used in modern wireless networks with significantly reduced EMF of UPWDs.

References

1 Boccardi, F., Heath, R.W., Lozano, A. et al. (2014). Five disruptive technology directions for 5G. *IEEE Communications Magazine* 52 (2): 74–80.

2 International Agency for Research on Cancer and others (2011). IARC classifies radiofrequency electromagnetic fields as possibly carcinogenic to humans. *Press Release* no. 208.

3 Jamshed, M.A., Heliot, F., and Brown, T. (2019). A survey on electromagnetic risk assessment and evaluation mechanism for future wireless communication systems. *IEEE Journal of Electromagnetics, RF and Microwaves in Medicine and Biology* 4 (1): 24–36.

4 Rehman, M.U., Gao, Y., Zhang, Z.W.J. et al. (2010). Investigation of on-body bluetooth transmission. *IET Microwaves, Antennas & Propagation* 4: 871–880.

5 Federal Communication Commission (2014). Specific Absorption Rate (SAR) for Cell Phones: What it Means for You.

6 Dai, L., Wang, B., Ding, Z. et al. (2018). A survey of non-orthogonal multiple access for 5G. *IEEE Communication Surveys and Tutorials* 20 (3): 2294–2323.

7 Khan, W.U., Jameel, F., Jamshed, M.A. et al. (2020). Efficient power allocation for noma-enabled IoT networks in 6G era. *Physical Communication* 39: 101043.

8 Jameel, F., Jamshed, M.A., Chang, Z. et al. (2020). Low latency ambient backscatter communications with deep Q-learning for beyond 5G applications. *2020 IEEE 91st Vehicular Technology Conference (VTC2020-Spring)*, pp. 1–6, IEEE.

9 Khan, W.U., Javed, M.A., Nguyen, T.N. et al. (2022). Energy-efficient resource allocation for 6G backscatter-enabled NOMA IoV networks. *IEEE Transactions on Intelligent Transportation Systems* 23 (7): 9775–9785.

10 Sambo, Y.A., Al-Imari, M., Héliot, F., and Imran, M.A. (2016). Electromagnetic emission-aware schedulers for the uplink of OFDM wireless communication systems. *IEEE Transactions on Vehicular Technology* 66 (2): 1313–1323.

11 Jamshed, M.A., Heliot, F., and Brown, T. (2021). Unsupervised learning based emission -aware uplink resource allocation scheme for non-orthogonal multiple access systems. *IEEE Transactions on Vehicular Technology* 70 (8): 7681–7691.

12 Velmurugan, T. and Santhanam, T. (2010). Computational complexity between K-means and K-medoids clustering algorithms for normal and uniform distributions of data points. *Journal of Computer Science* 6 (3): 363.

13 Golden, G.D., Foschini, C., Valenzuela, R.A., and Wolniansky, P.W. (1999). Detection algorithm and initial laboratory results using V-BLAST space-time communication architecture. *Electronics Letters* 35 (1): 14–16.

14 Ali, Z., Sidhu, G.A.S., Waqas, M., and Gao, F. (2018). On fair power optimization in nonorthogonal multiple access multiuser networks. *Transactions on Emerging Telecommunications Technologies* 29 (12): e3540.

15 Conil, E. (2013). D2.4 Global wireless exposure metric definition V1. *LexNet Project*.

16 Luo, Z.-Q. and Yu, W. (2006). An introduction to convex optimization for communications and signal processing. *IEEE Journal on Selected Areas in Communications* 24 (8): 1426–1438.

17 Al-Imari, M., Xiao, P., Imran, M.A., and Tafazolli, R. (2013). Low complexity subcarrier and power allocation algorithm for uplink OFDMA systems. *EURASIP Journal on Wireless Communications and Networking* 2013 (1): 1–6.

18 Wu, J., Rangan, S., and Zhang, H. (2016). *Green Communications: Theoretical Fundamentals, Algorithms, and Applications*. CRC Press.

10

Road Ahead for Low EMF User Proximity Devices

Muhammad Ali Jamshed[1], Fabien Héliot[2], Tim W.C. Brown[2], and Masood Ur Rehman[1]

[1]*James Watt School of Engineering, University of Glasgow, Glasgow, UK*
[2]*Institute of Communication Systems (ICS), Home of 5G and 6G Innovation Centre, University of Surrey, Guildford, UK*

10.1 Introduction

The EM field (EMF) exposure is generated by the transmission of electromagnetic (EM) waves, which are related with the usage of oscillating electrical power and various kinds of natural and man-made illumination. These EM waves are classified depending on their radioactive nature, which can be ionizing (extremely dangerous to humans since it causes cellular and DNA damage) or non-ionizing. Even before the discovery of X-rays in 1895 and Marconi's installation of the first comprehensive wireless system in 1897, the environment was exposed to radiations that were unidentified and emerging from experimental discharge tubes. With the introduction of broadcasting technologies (transmitting information in the non-ionizing region) in the early twentieth century (e.g. radio, television), the widespread use of EMF radiating devices in our daily lives (e.g. microwave oven), and, more recently, the development of cellular technology, exposure to EMF radiations has become more common. In future wireless systems, the quantity and kind of EMF exposure sources in the environment will significantly grow. Despite the lack of sufficient evidence about the short-term health effects of EMF exposure, EMF radiation from wireless devices has been categorized as potentially carcinogenic to humans (category B) by the international agency for research on cancer (IARC) and the world health organization (WHO) [1]. Meanwhile, the rise in EMF radiations in the 5G era may only enhance the potential long-term health concerns associated with EMF [2].

Low Electromagnetic Field Exposure Wireless Devices: Fundamentals and Recent Advances, First Edition.
Edited by Masood Ur Rehman and Muhammad Ali Jamshed.
© 2023 The Institute of Electrical and Electronics Engineers, Inc. Published 2023 by John Wiley & Sons, Inc.

10.2 Perception and Physiological Impact of EMF

10.2.1 Public's Perception of Exposure and Risk Assessment

Since wireless communication has become mainstream [3], researchers have worked hard to study the potential positive and negative consequences of EMF exposure on the human body. For example, the two European funded projects European health risk assessment network on electromagnetic fields exposure (EFHRAN) [4] and generalized EMF research using novel methods (GERoNIMO) [5] sought to identify and comprehend the hidden effects of EMF on health in order to improve the health risk assessment of EMF; EFHRAN, which began in 2009 and ended in 2012, relied on a network of broadband or frequency selective remote stations to measure and analyze EMF exposure [4]. The findings of EFHRAN revealed no compelling evidence relating exposure to EMF waves to any probable harmful effect on human health. In the meanwhile, GERoNIMO is a five-year initiative. In addition to refining the methods for assessing the health risks of EMF, this project intends to define the existing and future levels of EMF to which Europeans are exposed and, if required, to propose legislation to restrict EMF radiation.

In wireless communications, EMF exposure can be divided into two categories: active and passive. Active exposure refers to exposure generated from an intended usage of a wireless device, such as a mobile phone, laptop, or tablet used by a person to its own body/head, whereas passive exposure refers to ambient exposure, such as from access points (APs) or other people's wireless devices. Even though empirical studies (e.g. [2, 6, 7]) suggest that active exposure poses the greatest risk to health, because mobile device antenna(s) typically operate close to the user's head/body, most public concerns about EMF exposure are directed towards passive exposure; for example, two surveys from 2010 and 2013 revealed that the vast majority of people (70%) believe that mobile phone masts have an impact on their health [8] and APs are the primary sources of EMF exposure [9]. The most recurring queries are related to the EMF exposure levels within the vicinity of APs, effects of increased and indiscriminate deployment of APs and EMF regulatory compliance of AP antennas. The consequences of EMF radiation from APs located near humans and animals are still being contested, despite the fact that multiple reports [2, 6, 10, 11] indicate that this form of radiation is not capable of causing short-term health problems as long as there is no direct contact between the user and the antenna. While short-term health risks have been ruled out, long-term health impacts are currently under investigation [12].

10.2.2   Physiological Impact

Given that active exposure is considered more dangerous than passive exposure, most research on the physiological impact of EMF exposure in wireless communications have focused on active exposure [2]. A number of research have been conducted to investigate the link between the usage of mobile communication devices and an increased risk of brain cancer (e.g. glioma, meningioma, or acoustic neuroma) [2, 6, 7]. Recently, new studies have attempted to understand the differences in EMF radiation absorption in different age groups [13–15], as well as to investigate other possible effects on different organs/cognitive functions, such as the eye [13] or memory in the brain [16], and to investigate the effect of mmWave exposure on the human body [17].

10.2.2.1   Age Range and Exposure

According to current research, it appears that the way our bodies receive EMF radiations from mobile phones varies with age. There is currently sufficient data in the literature to prove that children receive a greater quantity of radiation than adults [14]. This is because differences in the dielectric characteristics of organic tissues affect EMF absorption, causing EMF radiation to permeate deeper into children's tissues than adult tissues. These findings were validated in [15] by utilizing finite-difference time-domain (FDTD) simulations; it was demonstrated that a child brain absorbs about 50% more radiation than an adult brain. Meanwhile, it was noted in [13] that it is not just children's brains that are at risk, but also their eyes, especially when youngsters use innovative programs like as virtual reality on their mobile phones. Another recent study in [16] shown that extended exposure to EMF radiation had a deleterious effect on adolescent figural memory ability. Figural memory is mostly located on the right side of the brain, and teenagers who used their phones on this side of the head saw a more dramatic loss in memory function.

10.2.2.2   mmWave and Exposure

The mmWave energy is insufficient to break hydrogen or ionic bonds, which are crucial in interactions between biological molecules within cells. However, mmWaves can cause certain free molecules with dipole moments to rotate, boosting body temperature [18]. The primary safety problem with the use of mmWave is the thermal biological effect of mmWave on the human body (particularly on the eyes and skin) produced by the absorption of EM mmWave energy by tissues, cells, and biological fluids [19, 20]. These thermal effects often manifest after being subjected to mmWave with an incident power density (IPD) greater than 5–10 mW/cm^2 [18]. High-intensity mmWaves have a dose-dependent effect on human skin and cornea: warm sensations can occur at low power densities, followed by discomfort at greater exposures, and potentially physical damage at extremely high powers. People who are exposed to radiation (high power at 94 GHz) have a rise in their surface temperature, according to [21]. Pain was observed to be associated with an increase in surface temperature in human volunteers [22]. It was also claimed in [23] that mmWave can have non-thermal biological effects (effects that occur at intensities much below any heating), given that these wavelengths approach the dimension of the skin and that the sweat duct serves as a helical antenna in the sub–THz band. Meanwhile, preliminary findings suggested that mmWave can alter gene expression, promote cellular proliferation and protein synthesis associated with oxidative stress, raise skin temperature, induce inflammatory and metabolic processes, cause ocular damage, and affect neuro-muscular dynamics [17]. It should be noted, however, that low intensity mmWaves have been proven beneficial in [24] to treat a variety of skin conditions and surface wounds.

10.2.2.3 Brain Tumour and Exposure

The significance of active exposure in the increased risk of brain tumor has ignited a long and contentious dispute in the scientific community, with studies yielding inconsistent results. According to the [25], the tissues that receive 80% of the radiation from a mobile phone, i.e. the tissues in direct touch with the head, may be most susceptible and prone to the effects of radio frequency (RF) radiation. Meanwhile, it has been argued in [2] that people who have used a cell phone for more than ten years are more likely to develop brain tumors. More recently, in [6], it was demonstrated that those who have heavily used a cellular phone for a period of time surpassing 25 years had an even greater risk of acquiring a brain tumor, prompting the recommendation that the present recommendations for exposure be quickly updated. In contrast, the largest-ever research on the issue, [26, 27], failed to discover compelling evidence that mobile phones enhance the incidence of brain tumors in 2010. Given the lack of general agreement on the long-term risks associated with the use of mobile devices, and more broadly, human exposure from EM sources, the WHO and IARC concluded that EMF radiation is possibly carcinogenic to humans and, as such, is classified as Group 2B [1]; a category used when there is limited evidence of carcinogenicity in humans. It is obvious that the academic community is still trying to reach a conclusion on the possible harm of EMF exposure from wireless communication systems to people after over 20 years of research. Outside of this community, two different Italian courts of justice [28] and [29] recently concluded that long-term exposure to EMF radiation from mobile phones may cause brain tumors (i.e. aligning their rulings with the conclusions of the scientific studies in [2, 6, 7]).

10.3 EMF Exposure Evaluation Metric and Regulations: A Future Perspective

10.3.1 Expected Exposure Contribution of Future Wireless Communication Technologies

The three major goals of the future generation of communication systems are to provide increased mobile broadband, huge machine-to-machine communication, and low latency communications [30]. 5G will rely on key enabling technologies such as densification, multiple-input multiple-output (MIMO), and mmWave to accomplish this. According to many recent works, [31–34], the use of densification, massive MIMO, and mmWave can help achieve the desirable peak data rate, spectral efficiency, area traffic capacity, latency, and so on that will be required to meet the objectives of 5G. However, these technologies will likely have an influence on EMF exposure, the magnitude of which has yet to be identified. To guard against documented harmful health effects, global harmonization of the RF EMF exposure limits for frequencies over 6 GHz is desirable, with a comparable margin of safety as for frequencies below 6 GHz. To the best of our knowledge, there are relatively few work in the literature relevant to the assessment of EMF exposure in 5G and future wireless devices. In following we describe some preliminary works that start to fill this knowledge gap.

10.3.1.1 Exposure and mmWave
According to the medical studies mentioned in this chapter, the main effect of mmWave on the body is an increase in temperature that can be hazardous in many different ways [18]. Meanwhile, the exposure measuring methodology for mmWave is not the same as that utilized for lower frequency RF waves. So far, work on monitoring mmWave exposure in 5G has been minimal, and it is evident that this is an area that needs to be researched more.

10.3.1.2 Exposure and Massive MIMO
The MIMO transceiver technology can be beneficial for decreasing overall transmit power by utilizing diversity, and so plays an important part in exposure reduction. When the channel state information is well known, it is generally known that transmit power decreases with the number of antennas [35]. Furthermore, it has been demonstrated in [36] that user devices equipped with numerous antennas can aid in reducing exposure. Beamforming is also possible with numerous antennas at the transmitter. Beamforming is yet another essential technology to reduce EMF exposure in 5G, by reducing the emitted power while maintaining the same signal-to-noise ratio at the receiver. In [37] and [38], beamforming-based methods are used to maximize overall system performance and minimize peak spatial

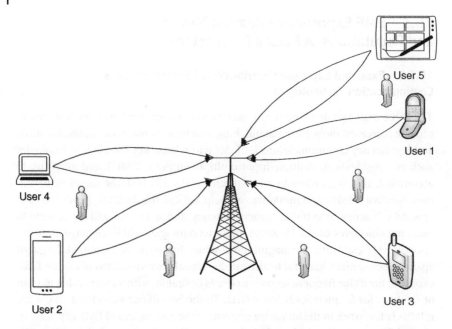

Figure 10.1 Exposure reduction scenario using MIMO antennas. Source: Jamshed et al. [39]/with permission of IEEE.

specific absorption rate (SAR) for multi-antenna wireless systems; the method in [37] can also be used to estimate the lower and upper bounds of the SAR within anatomically correct human head models, as well as to minimize the SAR within selected tissues in these models. Figure 10.1 demonstrates the usage of MIMO and beamforming in combination to decrease passive exposure in the downlink of a cellular system.

10.3.1.3 Exposure and Densification

Small cell deployment, as noted in [40], can assist to limit exposure; however, dense small cell deployment is still an area that has to be researched. For example, in the context of 5G, it was discovered in [41] that the maximum transmitted power over short distances when some antennas broadcast straight at the human body is much lower than the power levels employed by the third and fourth generation mobile devices. On the contrary, [42] claims that the aggregate RF field created in the downlink of a dense 5G deployment has much greater power than that generated by existing cellular systems.

10.3.2   Open Issues and Future Research Tracks

10.3.2.1   New EMF Limits and Guidelines

The main shortcomings of current EMF limitations have been clearly highlighted in [2] and [6], as well as by court rulings mentioned in [28] and [29], namely, that these limitations are designed for short-term usage and thus do not protect against the potential long-term effects of EMF exposure. As a result, [6] has pushed for a reform of these constraints in order to alleviate the rising public concern over EMF exposure from communications systems. These concerns are shown by certain new guidelines and/or legislation made lately in Europe that tacitly acknowledge the risk of prolonged exposure to EMF radiation. For example, the European Environmental Agency (EEA) suggested in 2013 preventative measures such as using a hands-free or earphone, minimizing the use of mobile phones (particularly in rapidly moving cars, i.e. under bad mobile channel circumstances), and restricting the use of mobile phones by youngsters [43]. Meanwhile, there are no specific guidelines regarding the maximum exposure levels on eyes when using mobile devices for video calling or virtual reality applications [13]; this is cause for concern given the increasing popularity of virtual reality applications, particularly among children.

10.3.2.2   EMF Mitigation Techniques and New Metrics

The exposure index (EI) proposed in [44], is an important step toward a unified and generic framework for measuring the exposure of wireless communication systems supporting a large geographical area and a large number of users. In that respect, it is highly beneficial for monitoring exposure within a geographically defined region, but not necessarily for limiting individual user exposure. Given that the possible harmful effects of exposure occur at the person level, we propose that greater efforts should be devoted toward the creation of a generic exposure measure for assessing individual exposure while accounting for active and passive exposures (i.e. uplink and downlink from diverse sources), as well as the thermal and non-thermal impacts of exposure. On the mitigation side, [45] and [46] are helpful beginning points for understanding the link between the SAR and the relative angle between antenna elements (i.e. it fluctuates as a function of the beam-pattern) in a mobile device with numerous antenna components. These works are based on a mobile device with two antennas and just one handling position (i.e. only one position of the mobile device relatives to the user head). It will be interesting to improve this model to include more antenna parts as well as more realistic mobile device handling locations based on current and future usages. Such models might then be used with re-configurable antennas to produce a more user-centric approach to EMF exposure; an approach in which a user is made aware of its exposure and can subsequently choose a preferred exposure profile (i.e. from low to high). The mobile device might then reorganize itself to fit the selected profile based on information of the device's relative position to the user's head or torso.

10.3.2.3 Other Open Issues

Based on the conversations, there are definitely some outstanding difficulties about the levels and evaluation of exposure in 5G and future wireless devices. For instance:

1. 5G networks will function in tandem with present mobile systems in the initial stage of 5G implementation, with an unavoidable worldwide rise in the exposure level that must be assessed.
2. Multiple antenna technology is a crucial enabler of 5G for reaching high data rates, however it will increase near-field exposure. For example, how many antennae may be installed on a consumer device while keeping a safe exposure level? Furthermore, the effect of the hand(s) location on a multi-antenna wireless device on the SAR must be studied.
3. Small cell deployment helps to minimize transmit power levels, but what about dense or very dense deployment of small cells? Will it boost aggregated power as compared with a macro cell and cause greater EMF exposure? This should be thoroughly investigated.
4. The number of smart Internet of things IoT devices in close proximity to humans is likely to skyrocket in the next years, and the impact of these devices on exposure must be assessed.
5. The existing RF exposure safety guidelines do not define limitations over 100 GHz, despite the fact that spectrum utilization will certainly shift to these bands over time. As a result, more research into the consequences of exposure at these frequencies [19] is required, followed by the establishment of new safety limits.

10.4 Conclusion

This chapter contains a comprehensive overview of exposure risk assessment. From the standpoint of human health, it appears that the likelihood of brain tumor remains the primary source of worry associated with the widespread use of wireless devices, despite the fact that the consequences of EMF exposure are now being studied in new sections of the body (e.g. eyes). Meanwhile, with the introduction of 5G, greater attempts are being made to comprehend the thermal and non-thermal impacts of mmWave radiation on the human body. We have offered some perspectives on how major 5G enabling technologies such as densification, MIMO, and mmWave may effect EMF exposure in the near future; for example, dense deployment of small cells and IoT devices would almost certainly increase total ambient exposure. We also feel that there may be some technological options in 5G to raise the exposure awareness of wireless system users and allow them to choose whether or not to lessen it at the expense of, say, a lower quality of service (QoS).

References

1 International Agency for Research on Cancer and others (2011). IARC classifies radio frequency electromagnetic fields as possibly carcinogenic to humans. *Press Release* no. 208.

2 Hardell, L.O., Carlberg, M., Söderqvist, F. et al. (2007). Long-term use of cellular phones and brain tumours-increased risk associated with use for greater than 10 years. *Occupational and Environmental Medicine* 64 (9): 626–632.

3 Dunnewijk, T. and Hultén, S. (2007). A brief history of mobile communication in Europe. *Telematics and Informatics* 24 (3): 164–179.

4 Elisabeth, C. (2012). EFHRAN-European health risk of assessment network on electromagnetic fields exposure. https://www.isglobal.org/en/-/efhran-european-health-risk-of-assessment-network-on-electromagnetic-fields-exposure (accessed 14 April 2019).

5 GERONIMO (2018). Generalized EMF research using novel methods. An integrated approach: from research to risk assessment and support to risk management. http://radiation.isglobal.org/index.php/en/project-description/geronimo-project-description (accessed 14 April 2019).

6 Hardell, L. and Carlberg, M. (2015). Mobile phone and cordless phone use and the risk for glioma analysis of pooled case-control studies in Sweden, 1997–2003 and 2007–2009. *Pathophysiology* 22 (1): 1–13.

7 Coureau, G., Bouvier, G., Lebailly, P. et al. (2014). Mobile phone use and brain tumours in the CERENAT case-control study. *Occupational and Environmental Medicine* 71 (7): 514–522.

8 EC (European Commission) (2007). Electromagnetic fields. Special Eurobarometer 272a/wave 66.2–TNS opinion and social.

9 Wiedemann, P.M. and Freudenstein, F. (2013). LEXNET - Low EMF Exposure Future Networks: D 2.2 Risk and Exposure Perception. *Tech. Rep.* Karlsruhe Institute of Technology ((KIT) / ITAS / WF-EMF).

10 Schmid, G., Preiner, P., Lager, D. et al. (2007). Exposure of the general public due to wireless LAN applications in public places. *Radiation Protection Dosimetry* 124 (1): 48–52.

11 Foster, K.R. (2007). Radiofrequency exposure from wireless LANs utilizing Wi-Fi technology. *Health Physics* 92 (3): 280–289.

12 Deniz, O.G., Kaplan, S., Selçuk, M.B. et al. (2017). Effects of short and long term electromagnetic fields exposure on the human hippo campus. *Journal of Microscopy and Ultrastructure* 5 (4): 191–197.

13 Fernández, C., de Salles, A.A., Sears, M. et al. (2018). Absorption of wireless radiation in the child versus adult brain and eye from cell phone conversation or virtual reality. *Environmental Research* 167: 694–699.

14 Morris, R.D., Morgan, L.L., and Davis, D. (2015). Children absorb higher doses of radio frequency electromagnetic radiation from mobile phones than adults. *IEEE Access* 3: 2379–2387.

15 Fernández-Rodríguez, C.E., De Salles, A.A.A., and Davis, D.L. (2015). Dosimetric simulations of brain absorption of mobile phone radiation–the relationship between psSAR and age. *IEEE Access* 3: 2425–2430.

16 Foerster, M., Thielens, A., Joseph, W. et al. (2018). A prospective cohort study of adolescents memory performance and individual brain dose of microwave radiation from wireless communication. *Environmental Health Perspectives* 126 (7): 1–13.

17 Di Ciaula, A. (2018). Towards 5G communication systems: are there health implications? *International Journal of Hygiene and Environmental Health* 221 (3): 367–375.

18 Le Dréan, Y., Mahamoud, Y.S., Le, Y. et al. (2013). State of knowledge on biological effects at 40–60 GHz. *Comptes Rendus Physique* 14 (5): 402–411.

19 Mumtaz, S., Rodriguez, J., and Dai, L. (2017). Introduction to mmWave massive MIMO. In: *mmWave Massive MIMO*, 1–18, Elsevier.

20 Wu, T., Rappaport, T.S., and Collins, C.M. (2015). Safe for generations to come: considerations of safety for millimeter waves in wireless communications. *IEEE Microwave Magazine* 16 (2): 65–84.

21 Debouzy, J., Crouzier, D., Dabouis, V. et al. (2007). Biologic effects of millimeteric waves (94 GHz). Are there long term consequences? *Pathologie-Biologie* 55 (5): 246–255.

22 Walters, T.J., Blick, D.W., Johnson, L.R. et al. (2000). Heating and pain sensation produced in human skin by millimeter waves: comparison to a simple thermal model. *Health Physics* 78 (3): 259–267.

23 Betzalel, N., Ishai, P.B., and Feldman, Y. (2018). The human skin as a sub-THz receiver–does 5G pose a danger to it or not? *Environmental Research* 163: 208–216.

24 Ziskin, M.C. (2013). Millimeter waves: acoustic and electromagnetic. *Bioelectromagnetics* 34 (1): 3–14.

25 Cardis, E., Deltour, I., Mann, S. et al. (2008). Distribution of RF energy emitted by mobile phones in anatomical structures of the brain. *Physics in Medicine and Biology* 53 (11): 2771–2783.

26 INTERPHONE Study Group (2010). Brain tumour risk in relation to mobile telephone use: results of the INTERPHONE international case–control study. *International Journal of Epidemiology* 39 (3): 675–694.

27 Wild, C. (2011). IARC Report to the Union for International Cancer Control (UICC) on the Interphone Study.

28 Osborn, A. (2012). Italy court ruling links mobile phone use to tumor. https://www.reuters.com/article/us-italy-phones/italy-court-ruling-links-mobile-phone-use-to-tumor-idUSBRE89I0V320121019 (accessed 15 November 2018).

29 France-Presse, A. (2017). Italian court rules mobile phone use caused brain tumour. https://www.theguardian.com/technology/2017/apr/21/italian-court-rules-mobile-phone-use-caused-brain-tumour (accessed 15 November 2018).

30 Series, M. (2015). IMT vision–framework and overall objectives of the future development of IMT for 2020 and beyond. *Recommendation ITU*, 1–19.

31 Vahid, S., Tafazolli, R., and Filo, M. (2015). *Small Cells for 5G Mobile Networks*, Chapter 3, 63–104. Wiley-Blackwell.

32 Elijah, O., Leow, C.Y., Rahman, T.A. et al. (2016). A comprehensive survey of pilot contamination in massive MIMO 5G system. *IEEE Communication Surveys and Tutorials* 18 (2): 905–923.

33 de Carvalho, E., Popovski, P., Thomsen, H. et al. (2013). EU FP7 INFSO-ICT-317669 METIS, D3.1 Positioning of Multi-Node/Multi-Antenna Technologies. *Tech. Rep.* EU FP7 INFSO-ICT-317669-METIS.

34 Rappaport, T.S., Sun, S., Mayzus, R. et al. (2013). Millimeter wave mobile communications for 5G cellular: it will work!. *IEEE Access* 1 (1): 335–349.

35 Ngo, H.Q., Larsson, E.G., and Marzetta, T.L. (2011). Uplink power efficiency of multiuser MIMO with very large antenna arrays. *2011 49th Annual Allerton Conference on Communication, Control, and Computing (Allerton)*, pp. 1272–1279, IEEE.

36 Baldauf, M.A., Pontes, J.A., Timmermann, J., and Wiesbeck, W. (2007). Mobile MIMO phones and their human exposure to electromagnetic fields. *International Conference on Electromagnetics in Advanced Applications, 2007. ICEAA 2007*, pp. 9–12, IEEE.

37 Wang, M., Lin, L., Chen, J. et al. (2011). Evaluation and optimization of the specific absorption rate for multiantenna systems. *IEEE Transactions on Electromagnetic Compatibility* 53 (3): 628–637.

38 Ying, D., Love, D.J., and Hochwald, B.M. (2013). Beamformer optimization with a constraint on user electromagnetic radiation exposure. *2013 47th Annual Conference on Information Sciences and Systems (CISS)*, pp. 1–6, IEEE.

39 Jamshed, M.A., Heliot, F., and Brown, T.W. (2019). A survey on electromagnetic risk assessment and evaluation mechanism for future wireless communication systems. *IEEE Journal of Electromagnetics, RF and Microwaves in Medicine and Biology* 4 (1): 24–36.

40 Aerts, S., Plets, D., Verloock, L. et al. (2013). Assessment and comparison of total RF-EMF exposure in femtocell and macrocell base station scenarios. *Radiation Protection Dosimetry* 162 (3): 236–243.

41 Thors, B., Colombi, D., Ying, Z. et al. (2016). Exposure to RF EMF from array antennas in 5G mobile communication equipment. *IEEE Access* 4: 7469–7478.

42 Nasim, I. and Kim, S. (2017). Human Exposure to RF Fields in 5G Downlink. *arXiv preprint arXiv:1711.03683.*

43 Hardell, L., Carlberg, M., and Gee, D. (2012). 21 mobile phone use and brain tumour risk: early warnings, early actions? *Late Lessons from Early Warnings: Science, Precaution, Innovation*, pp. 509–529.

44 Varsier, N., Plets, D., Corre, Y. et al. (2015). A novel method to assess human population exposure induced by a wireless cellular network. *Bioelectromagnetics* 36 (6): 451–463.

45 Chim, K.-C., Chan, K.C., and Murch, R.D. (2004). Investigating the impact of smart antennas on SAR. *IEEE Transactions on Antennas and Propagation* 52 (5): 1370–1374.

46 Hochwald, B.M., Love, D.J., Yan, S., and Jin, J. (2013). SAR codes. *Information Theory and Applications Workshop (ITA), 2013*, pp. 1–9, IEEE.

Index

Low Electromagnetic Field Exposure Wireless Devices: Fundamentals and Recent Advances, First Edition.
Edited by Masood Ur Rehman and Muhammad Ali Jamshed.
© 2023 The Institute of Electrical and Electronics Engineers, Inc. Published 2023 by John Wiley & Sons, Inc.

Printed and bound by CPI Group (UK) Ltd, Croydon, CR0 4YY

16/04/2025